Pro Data Visualization Using R and JavaScript

Analyze and Visualize Key Data on the Web

Second Edition

Tom Barker
Jon Westfall

Apress®

Pro Data Visualization Using R and JavaScript: Analyze and Visualize Key Data on the Web

Tom Barker
Pipersville, PA, USA

Jon Westfall
Cleveland, MS, USA

ISBN-13 (pbk): 978-1-4842-7201-5
https://doi.org/10.1007/978-1-4842-7202-2

ISBN-13 (electronic): 978-1-4842-7202-2

Managing Director, Apress Media LLC: Welmoed Spahr
Acquisitions Editor: Steve Anglin
Development Editor: Matthew Moodie
Coordinating Editor: Mark Powers

Cover designed by eStudioCalamar

Cover image by Zanwei Guo on Unsplash (www.unsplash.com)

Distributed to the book trade worldwide by Apress Media, LLC, 1 New York Plaza, New York, NY 10004, U.S.A. Phone 1-800-SPRINGER, fax (201) 348-4505, e-mail orders-ny@springer-sbm.com, or visit www.springeronline. com. Apress Media, LLC is a California LLC and the sole member (owner) is Springer Science + Business Media Finance Inc (SSBM Finance Inc). SSBM Finance Inc is a **Delaware** corporation.

For information on translations, please e-mail booktranslations@springernature.com; for reprint, paperback, or audio rights, please e-mail bookpermissions@springernature.com.

Apress titles may be purchased in bulk for academic, corporate, or promotional use. eBook versions and licenses are also available for most titles. For more information, reference our Print and eBook Bulk Sales web page at http://www.apress.com/bulk-sales.

Any source code or other supplementary material referenced by the author in this book is available to readers on GitHub via the book's product page, located at www.apress.com/9781484272015. For more detailed information, please visit http://www.apress.com/source-code.

Printed on acid-free paper

For my grandmother, Ann Biango, who passed away during the creation of this book. I was very lucky to have her in my life for as long as I did.

—Tom Barker

Table of Contents

About the Authors

Tom Barker is the Senior Manager of Web Development at Comcast. He has authored *Pro JavaScript Performance: Monitoring and Visualization* and co-authored *Foundation Website Creation with HTML5, CSS3, and JavaScript*. Tom has also served as an adjunct professor at Philadelphia University for the last ten years. He lives outside of Philadelphia with his wife and two children.

Jon Westfall is an associate professor of psychology at Delta State University. He has authored *Set Up and Manage Your Virtual Private Server, Practical R 4, Beginning Android Web Apps Development, Windows Phone 7 Made Simple*, and several works of fiction including *One in the Same, Mandate, and Franklin: The Ghost Who Successfully Evicted Hipsters from His Home and Other Short Stories*. He lives in Cleveland, Mississippi, with his wife.

About the Technical Reviewer

 **Matt Wiley** leads institutional effectiveness, research, and assessment at Victoria College, facilitating strategic and unit planning, data-informed decision making, and state/regional/federal accountability. As a tenured, associate professor of mathematics, he won awards in both mathematics education (California) and student engagement (Texas). Matt holds degrees in computer science, business, and pure mathematics from the University of California and Texas A&M systems.

Outside academia, he has co-authored three books about the popular R programming language and was managing partner of a statistical consultancy for almost a decade. His programming experience is with R, SQL, C++, Ruby, Fortran, and JavaScript.

A programmer, a published author, a mathematician, and a transformational leader, Matt has always melded his passion for writing with his joy of logical problem solving and data science. From the boardroom to the classroom, he enjoys finding dynamic ways to partner with interdisciplinary and diverse teams to make complex ideas and projects understandable and solvable. Matt enjoys being found online via Twitter at @matt math or `http://mattwiley.org/`.

Acknowledgments

I want to thank Ben Renow-Clarke for thinking of me for this great project. I want to thank Matthew Moodie and Christine Rickets and the rest of the team at Apress for their guidance and direction. I want to thank Matt Canning for helping me see the code with fresh eyes and for keeping me honest.

I want to thank my team at Comcast: every one of you is amazing and I am made better by being a part of such an incredible team.

I want to thank my amazing wife Lynn and our beautiful children Lukas and Paloma for their patience and understanding while I would write every night until late in the night.

—Tom Barker

I'd like to thank my wife and parents for their love and support. I'd also like to thank the team at Apress for their hard work to help make this project a reality.

—Jon Westfall

CHAPTER 1

Background

When the first edition of this text was released, there was a new concept emerging in the field of web development: using data visualizations as communication tools. Today, Infographics are everywhere on the Net; however, this concept is something that was already well established in other fields and departments for generations. At the company where you work, your finance department probably uses data visualizations to represent fiscal information both internally and externally; just take a look at the quarterly earnings reports for almost any publicly traded company. They are full of charts to show revenue by quarter, or year over year earnings, or a plethora of other historic financial data. All are designed to show lots and lots of data points, potentially pages and pages of data points, in a single easily digestible graphic.

Compare the bar chart in Google's quarterly earnings report from back in 2007 (ah, when Google was a "small" company; see Figure 1-1) to a subset of the data it is based on in tabular format (see Figure 1-2).

© Tom Barker, Jon Westfall 2022
T. Barker and J. Westfall, *Pro Data Visualization Using R and JavaScript*,
https://doi.org/10.1007/978-1-4842-7202-2_1

Quarterly Revenue

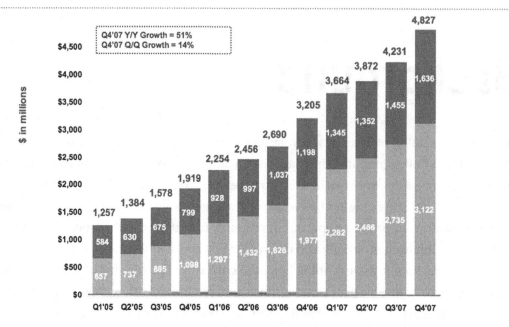

Figure 1-1. *Google Q4 2007 quarterly revenue shown in a bar chart*

	Class A and Class B Common Stock		Additional Paid-In Capital	Deferred Stock Based Compensation	Accumulated Other Comprehensive Income	Retained Earnings	Total Stockholders' Equity
	Shares	Amount	Amount				
Balance at January 1, 2005	266,917	$ 267	$ 2,582,352	$ (249,470)	$ 5,436	$ 590,471	$ 2,929,056
Issuance of common stock in connection with follow-on public offering and acquisitions, net	14,869	15	4,316,022	(2,036)	—	—	4,314,001
Stock-based award activity	11,241	11	579,418	132,491	—	—	711,920
Comprehensive income:							
Change in unrealized gain (loss) on available-for-sale investments, net of tax effect of $11,404	—	—	—	—	16,580	—	16,580
Foreign currency translation adjustment	—	—	—	—	(17,997)	—	(17,997)
Net income	—	—	—	—	—	1,465,397	1,465,397
Total comprehensive income	—	—	—	—	—	—	1,463,980
Balance at December 31, 2005	293,027	293	7,477,792	(119,015)	4,019	2,055,868	9,418,957
Issuance of common stock in connection with follow-on public offering and acquisitions, net	7,689	8	3,236,778	—	—	—	3,236,786
Stock-based award activity	8,281	8	1,168,336	119,015	—	—	1,287,359
Comprehensive income:							
Change in unrealized gain (loss) on available-for-sale investments, net of tax effect of $13,280	—	—	—	—	(19,309)	—	(19,309)
Foreign currency translation adjustment	—	—	—	—	38,601	—	38,601
Net income	—	—	—	—	—	3,077,446	3,077,446
Total comprehensive income	—	—	—	—	—	—	3,096,738
Balance at December 31, 2006	308,997	309	11,882,906	—	23,311	5,133,314	17,039,840
Stock-based award activity	3,920	4	1,358,315	—	—	—	1,358,319
Comprehensive income:							
Change in unrealized gain (loss) on available-for-sale investments, net of tax effect of $19,963	—	—	—	—	29,029	—	29,029
Foreign currency translation adjustment	—	—	—	—	61,033	—	61,033
Net income	—	—	—	—	—	4,203,720	4,203,720
Total comprehensive income	—	—	—	—	—	—	4,293,782
Adjustment to retained earnings upon adoption of FIN 48	—	—	—	—	—	(2,262)	(2,262)
Balance at December 31, 2007	312,917	$ 313	$ 13,241,221	$ —	$ 113,373	$ 9,334,772	$ 22,689,679

Figure 1-2. *Similar earnings data in tabular form*

The bar chart is imminently more readable. We can clearly see by the shape of it that earnings are up and have been steadily going up each quarter. By the color coding, we can see the sources of the earnings, and with the annotations, we can see both the precise numbers that those color coding represent and what the year over year percentages are.

With the tabular data, you have to read labels on the left, line up the data on the right with those labels, do your own aggregation and comparison, and draw your own conclusions. There is a lot more upfront work needed to take in the tabular data, and there exists the very real possibility of your audience either not understanding the data (thus creating their own incorrect story around the data) or tuning out completely because of the sheer amount of work needed to take in the information.

It's not just the finance department that uses visualizations to communicate dense amounts of data. Maybe your operations department uses charts to communicate server uptime, or your customer support department uses graphs to show call volume. Whatever the case, it's no wonder that engineering and web development groups have finally gotten on board with this.

As part of any department, group, or industry, we have a huge amount of relevant data that is important for us to first be aware of so that we can refine and improve what we do, but also to communicate out to our stakeholders, to demonstrate our successes or validate resource needs, or to plan tactical roadmaps for the coming year.

Before we can do this, we need to understand what we are doing. We need to understand what data visualizations are, a general idea of their history, when to use them, and how to use them both technically and ethically.

What Is Data Visualization?

OK, so what exactly is data visualization? Data visualization is the art and practice of gathering, analyzing, and graphically representing empirical information. They are sometimes called *information graphics* ("Infographics"), or even just *charts* and *graphs*. Whatever you call it, the goal of visualizing data is to tell the story in the data. Telling the story is predicated on understanding the data at a very deep level and gathering insight from comparisons of data points in the numbers.

There exists syntax for crafting data visualizations, patterns in the form of charts that have an immediately known context. We devote a chapter to each of the significant chart types later in the book.

Time Series Charts

Time series charts show changes over time. See Figure 1-3 for a time series chart that shows the weighted popularity of the keyword "Data Visualization" from Google Trends (`www.google.com/trends/`).

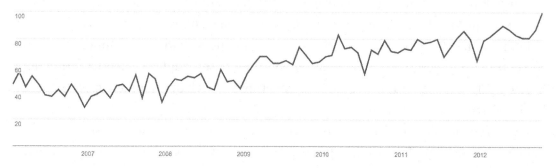

Figure 1-3. *Time series of weighted trend for the keyword "Data Visualization" from Google Trends*

Note that the vertical y-axis shows a sequence of numbers that increment by 20 up to 100. These numbers represent the weighted search volume, where 100 is the peak search volume for our term. On the horizontal x-axis, we see years going from 2007 to 2012. The line in the chart represents both axes, the given search volume for each date.

From just this small sample size, we can see that the term has more than tripled in popularity, from a low of 29 in the beginning of 2007 up to the ceiling of 100 by the end of 2012.

Bar Charts

Bar charts show comparisons of data points. See Figure 1-4 for a bar chart that demonstrates the search volume by country for the keyword "Data Visualization," the data for which is also sourced from Google Trends.

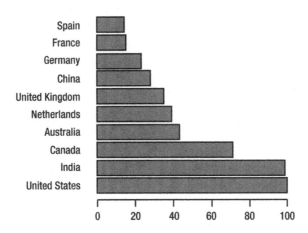

**Search Volume for Keyword
'Data Visualization' by Region
from Google Trends**

Figure 1-4. *Google Trends breakdown of search volume by region for keyword "Data Visualization"*

We can see the names of the countries on the y-axis and the normalized search volume, from 0 to 100, on the x-axis. Notice, though, that no time measure is given. Does this chart represent data for a day, a month, or a year?

Also note that we have no context for what the unit of measure is. I highlight these points not to answer them but to demonstrate the limitations and pitfalls of this particular chart type. We must always be aware that our audience does not bring the same experience and context that we bring, so we must strive to make the stories in our visualizations as self-evident as possible.

Histograms

Histograms are a type of bar chart that displays continuous data on both axes. It is used to show the distribution of data or how often groups of information appear in the data. See Figure 1-5 for a histogram that shows how many articles the *New York Times* published each year, from 1980 to 2012, that related in some way to the subject of data visualization. We can see from the chart that the subject has been ramping up in frequency since 2009.

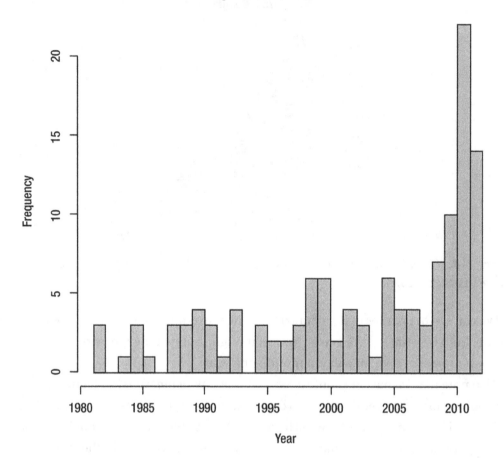

Figure 1-5. *Histogram showing distribution of NY Times articles about data visualization*

Data Maps

Data maps are used to show the distribution of information over a spatial region. Figure 1-6 shows a data map used to demonstrate the interest in the search term "Data Visualization" broken out by US states.

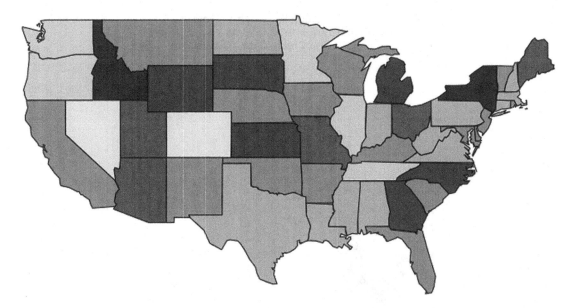

Figure 1-6. *Data map of US states by interest in "Data Visualization" (data from Google Trends)*

In this example, the states with the darker shades indicate a greater interest in the search term. (This data also is derived from Google Trends, for which interest is demonstrated by how frequently the term "Data Visualization" is searched for on Google.) It's also worth noting that while darker shades tend to be used to indicate greater impact, without a legend, we wouldn't know this for sure.

Scatter Plots

Like bar charts, *scatter plots* are used to compare data, but specifically to suggest correlations in the data, or where the data may be dependent or related in some way. See Figure 1-7, in which we use data from Google Correlate (`www.google.com/trends/correlate`), to look for a relationship between search volume for the keyword "What is Data Visualization" and the keyword "How to Create Data Visualization."

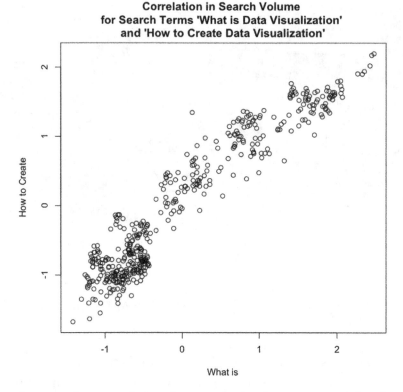

Figure 1-7. *Scatter plot examining the correlation between search volume for terms related to "Data Visualization," "How to Create," and "What is"*

This chart suggests a positive correlation in the data, meaning that as one term rises in popularity, the other also rises. So what this chart suggests is that as more people find out about data visualization, more people want to learn how to create data visualizations.

The important thing to remember about correlation is that it does not suggest a direct cause—correlation is not causation. Just because two numbers move in the same direction, does not mean one is causing the other to change. There could always be a third variable, or coincidence, causing the correlation.

History

If we're talking about the history of data visualization, the modern conception of data visualization largely started with William Playfair. William Playfair was, among other things, an engineer, an accountant, a banker, and an all-around Renaissance man who single-handedly created the time series chart, the bar chart, and the bubble chart. Playfair's charts were published in the late eighteenth century into the early nineteenth century. He was very aware that his innovations were the first of their kind, at least in the realm of communicating statistical information, and he spent a good amount of space in his books describing how to make the mental leap to seeing bars and lines as representing physical things like money.

Playfair is best known for two of his books: the *Commercial and Political Atlas* and the *Statistical Breviary*. The *Commercial and Political Atlas* was published in 1786 and focused on different aspects of economic data from national debt to trade figures and even military spending. It also featured the first printed time series graph and bar chart.

His *Statistical Breviary* focused on statistical information around the resources of the major European countries of the time and introduced the bubble chart.

Playfair had several goals with his charts, among them perhaps stirring controversy, commenting on the diminishing spending power of the working class, and even demonstrating the balance of favor in the import and export figures of the British Empire, but ultimately his most wide-reaching goal was to communicate complex statistical information in an easily digested, universally understood format.

Note   Both books are back in print relatively recently, thanks to Howard Wainer, Ian Spence, and Cambridge University Press.

Playfair had several contemporaries, including Dr. John Snow, who made my personal favorite chart: the cholera map. The cholera map is everything an informational graphic should be: it was simple to read, it was informative, and, most importantly, it solved a real problem.

The cholera map is a data map that outlined the location of all the diagnosed cases of cholera in the outbreak of London 1854 (see Figure 1-8). The shaded areas are recorded deaths from cholera, and the shaded circles on the map are water pumps. From careful inspection, the recorded deaths seemed to radiate out from the water pump on Broad Street.

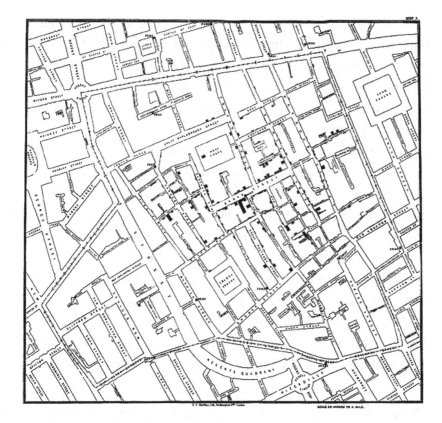

Figure 1-8. *John Snow's cholera map*

Dr. Snow had the Broad Street water pump closed, and the outbreak ended. Beautiful, concise, and logical.

Another historically significant information graphic is the Diagram of the Causes of Mortality in the Army in the East, by Florence Nightingale and William Farr. This chart is shown in Figure 1-9.

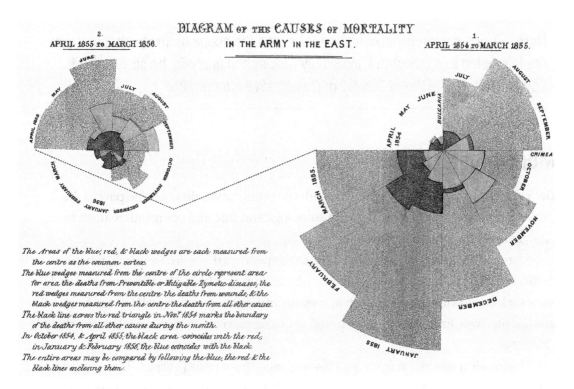

Figure 1-9. *Florence Nightingale and William Farr's Diagram of the Causes of Mortality in the Army in the East*

Nightingale and Farr created this chart in 1856 to demonstrate the relative number of preventable deaths and, at a higher level, to improve the sanitary conditions of military installations. Note that the Nightingale and Farr visualization is a stylized pie chart. Pie charts are generally a circle representing the entirety of a given data set with slices of the circle representing percentages of a whole. The usefulness of pie charts is sometimes debated because it can be argued that it is harder to discern the difference in value between angles than it is to determine the length of a bar or the placement of a line against Cartesian coordinates. Nightingale seemingly avoids this pitfall by having not just the angle of the wedge hold value but by also altering the relative size of the slices so they eschew the confines of the containing circle and represent relative value. This likely wins over some of the detractors of pie charts; however, in some circles of science and academia, there is no such thing as a good pie chart!

All the above examples had specific goals or problems that they were trying to solve.

> **Note**   A rich comprehensive history is beyond the scope of this book, but if you
> are interested in a thoughtful, incredibly researched analysis, be sure to read
> Edward Tufte's *The Visual Display of Quantitative Information.*

Modern Landscape

Data visualization is in the midst of a modern revitalization due in large part to the proliferation of cheap storage space to store logs and free and open source tools to analyze and chart the information in these logs.

From a consumption and appreciation perspective, there are websites that are dedicated to studying and talking about information graphics. There are generalized sites such as FlowingData that both aggregate and discuss data visualizations from around the Web, from astrophysics timelines to mock visualizations used on the floor of Congress.

The mission statement from the FlowingData About page (`http://flowingdata.com/about/`) is appropriately the following: "FlowingData explores how designers, statisticians, and computer scientists use data to understand ourselves better—mainly through data visualization."

There are more specialized sites such as quantifiedself.com that are focused on gathering and visualizing information about oneself. There are even web comics about data visualization, the quintessential one being xkcd.com, run by Randall Munroe. One of the most famous and topical visualizations that Randall has created thus far is the Radiation Dose Chart. We can see the Radiation Dose Chart in Figure 1-10 (it is available in high resolution here: `http://xkcd.com/radiation/`).

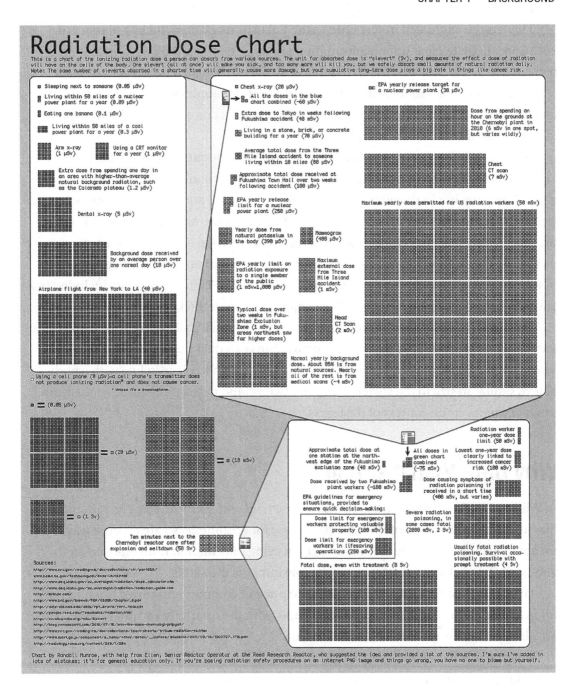

Figure 1-10. *Radiation Dose Chart, by Randall Munroe. Note that the range in scale being represented in this visualization as a single block in one chart is exploded to show an entirely new microcosm of context and information. This pattern is repeated over and over again to show an incredible depth of information*

This chart was created in response to the Fukushima Daiichi nuclear disaster of 2011 and sought to clear up misinformation and misunderstanding of comparisons being made around the disaster. It did this by demonstrating the differences in scale for the amount of radiation from sources such as other people or a banana up to what a fatal dose of radiation ultimately would be—how all that compared to spending just ten minutes near the Chernobyl meltdown.

Over the last quarter of a century, Edward Tufte, author and professor emeritus at Yale University, has been working to raise the bar of information graphics. He published groundbreaking books detailing the history of data visualization, tracing its roots even further back than Playfair to the beginnings of cartography. Among his principles is the idea to maximize the amount of information included in each graphic—both by increasing the amount of variables or data points in a chart and by eliminating the use of what he has coined chartjunk. *Chartjunk*, according to Tufte, is anything included in a graph that is not information, including ornamentation or thick, gaudy arrows.

Tufte also invented the *sparkline*, a time series chart with all axes removed and only the trend line remaining to show historic variations of a data point without concern for exact context. Sparklines are intended to be small enough to place in line with a body of text, similar in size to the surrounding characters, and to show the recent or historic trend of whatever the context of the text is.

Why Data Visualization?

In William Playfair's introduction to the *Commercial and Political Atlas*, he rationalizes that just as algebra is the abbreviated shorthand for arithmetic, so are charts a way to "abbreviate and facilitate the modes of conveying information from one person to another." Almost 300 years later, this principle remains the same.

Data visualizations are a universal way to present complex and varied amounts of information, as we saw in our opening example with the quarterly earnings report. They are also powerful ways to tell a story with data.

Imagine you have your Apache logs in front of you, with thousands of lines all resembling the following:

```
127.0.0.1 - - [10/Dec/2012:10:39:11 +0300] "GET / HTTP/1.1" 200 468 "-"
"Mozilla/5.0 (X11; U; Linux i686; en-US; rv:1.8.1.3) Gecko/20061201
Firefox/2.0.0.3 (Ubuntu-feisty)"
```

```
127.0.0.1 - - [10/Dec/2012:10:39:11 +0300] "GET /favicon.ico HTTP/1.1" 200
766 "-" "Mozilla/5.0 (X11; U; Linux i686; en-US; rv:1.8.1.3) Gecko/20061201
Firefox/2.0.0.3 (Ubuntu-feisty)"
```

Among other things, we see IP address, date, requested resource, and client user agent. Now imagine this repeated thousands of times—so many times that your eyes kind of glaze over because each line so closely resembles the ones around it that it's hard to discern where each line ends, let alone what cumulative trends exist within.

By using some analysis and visualization tools such as R, or even a commercial product such as Splunk, we can artfully pull out all kinds of meaningful and interesting stories out of this log, from how often certain HTTP errors occur and for which resources to what our most widely used URLs are, to what the geographic distribution of our user base is.

This is just our Apache access log. Imagine casting a wider net, pulling in release information, bugs, and production incidents. What insights we could gather about what we do: from how our velocity impacts our defect density to how our bugs are distributed across our feature sets. And what better way to communicate those findings and tell those stories than through a universally digestible medium, like data visualizations?

The point of this book is to explore how we as developers can leverage this practice and medium as part of continual improvement—both to identify and quantify our successes and opportunities for improvements and more effectively communicate our learning and our progress.

Tools

There are a number of excellent tools, environments, and libraries that we can use both to analyze and visualize our data. The next two sections describe them.

Languages, Environments, and Libraries

The tools that are most relevant to web developers are Splunk, R, and the D3 (Data-Driven Documents) JavaScript library. See Figure 1-11 for a comparison of interest over time for them (from Google Trends).

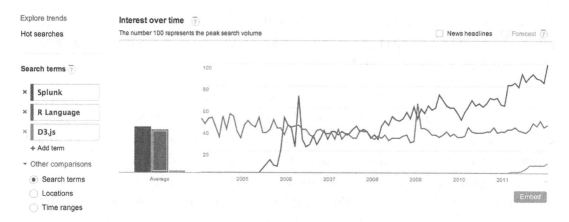

Figure 1-11. *Google Trends analysis of interest over time in Splunk, R, and D3*

From the figure, we can see that R has had a steady consistent amount of interest since 200; Splunk had an introduction to the chart around 2005, had a spike of interest around 2006, and had steady growth since then, which only started tapering off in 2019. As for D3, we see it just start to peak around 2011 when it was introduced and its predecessor Protovis was sunsetted. R and D3 have remained relatively stable in interest in the years since 2013.

Let's start with the tool of choice for many developers, scientists, and statisticians: the R language. We have a deep dive into the R environment and language in the next chapter, but for now, it's enough to know that it is an open source environment and language used for statistical analysis and graphical display. It is powerful, fun to use, and, best of all, it is free.

Splunk has seen a tremendous steady growth in interest over the last few years—and for good reason. It is easy to use once it's set up, scales wonderfully, supports multiple concurrent users, and puts data reporting at the fingertips of everyone. You simply set it up to consume your log files; then you can go into the Splunk dashboard and run reports on key values within those logs. Splunk creates visualizations as part of its reporting capabilities, as well as alerting. While Splunk is a commercial product, it also offers a free trial version, available here: `www.splunk.com/download`.

D3 is a JavaScript library that allows us to craft interactive visualizations. It is the official follow-up to Protovis. Protovis was a JavaScript library created in 2009 by Stanford University's Stanford Visualization Group. Protovis was sunsetted in 2011, and the creators unveiled D3. We explore the D3 library at length in Chapter 4.

Analysis Tools

Aside from the previously mentioned languages and environments, there are a number of analysis tools available online.

A great hosted tool for analysis and research is Google Trends. Google Trends allows you to compare trends on search terms. It provides all kinds of great statistical information around those trends, including comparing their relative search volume (see Figure 1-12), the geographic area those trends are coming from (see Figure 1-13), and related keywords.

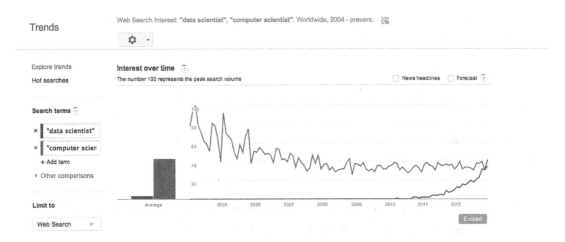

Figure 1-12. *Google Trends for the terms "data scientist" and "computer scientist" over time; note the interest in the term "data scientist" growing rapidly from 2011 on to match the interest in the term "computer scientist"*

Figure 1-13. *Google Trends data map showing geographic location where interest in the keywords is originating*

Another great tool for analysis is Wolfram|Alpha (http://wolframalpha.com). See Figure 1-14 for a screenshot of the Wolfram|Alpha home page.

Figure 1-14. *Home page for Wolfram|Alpha*

Wolfram|Alpha is not a search engine. Search engines spider and index content. Wolfram|Alpha is instead a question answering (QA) engine that parses human-readable sentences with natural language processing and responds with computed results. Say, for example, you want to search for the speed of light. You might go to the Wolfram|Alpha site and type in "what is the speed of light". Remember that it uses natural language processing to parse your search query, not the keyword lookup.

The results of this query can be seen in Figure 1-15. Wolfram|Alpha essentially looks up all the data it has around the speed of light and presents it in a structured, categorized fashion. You can also export the raw data for each result.

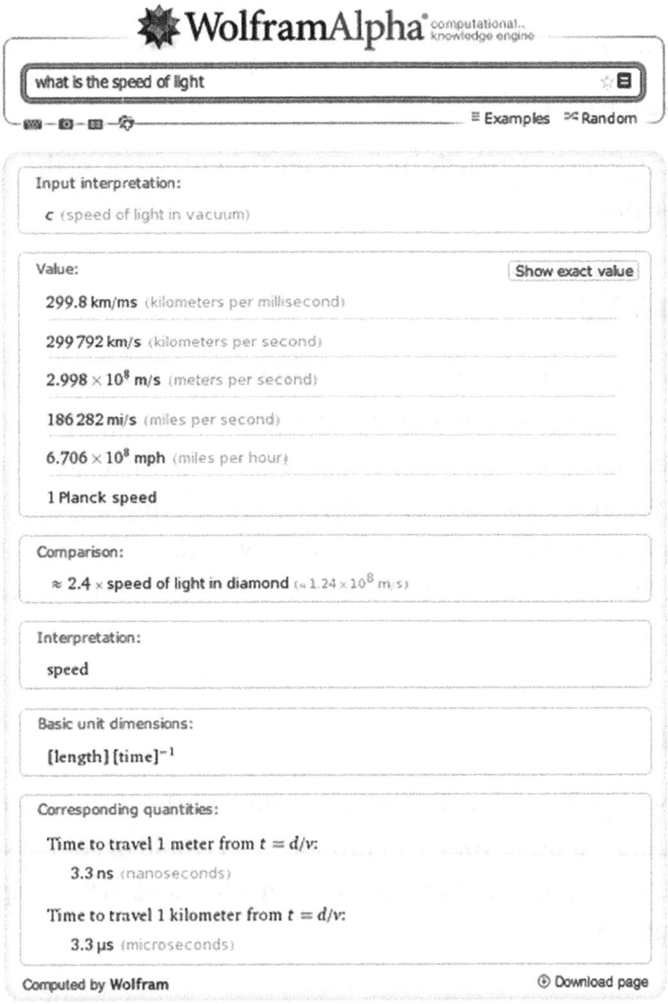

Figure 1-15. *Wolfram|Alpha results for query "what is the speed of light"*

Process Overview

So we understand what data visualization is and have a high-level understanding of the history of it and an idea of the current landscape. We're beginning to get an inkling about how we can start to use this in our world. We know some of the tools that are available to us to facilitate the analysis and creation of our charts. Now let's look at the process involved.

Creating data visualizations involves four core steps:

1. Identify a problem.

2. Gather the data.

3. Analyze the data.

4. Visualize the data.

Let's walk through each step in the process and re-create one of the previous charts to demonstrate the process.

Identify a Problem

The very first step is to identify a problem we want to solve. This can be almost anything—from something as profound and wide reaching as figuring out why your bug backlog doesn't seem to go down and stay down to seeing what feature releases over a given period in time caused the most production incidents and why.

For our example, let's re-create Figure 1-5 and try to quantify the interest in data visualization over time as represented by the number of *New York Times* articles on the subject.

Gather Data

We have an idea of what we want to investigate, so let's dig in. If you are trying to solve a problem or tell a story around your own product, you would of course start with your own data—maybe your Apache logs, maybe your bug backlog, maybe exports from your project tracking software.

Note If you are focusing on gathering metrics around your product and you don't already have data handy, you need to invest in instrumentation. There are many ways to do this, usually by putting logging in your code. At the very least, you want to log error states and monitor those, but you may want to expand the scope of what you track to include for debugging purposes while still respecting both your user's privacy and your company's privacy policy. In my book, *Pro JavaScript Performance: Monitoring and Visualization*, I explore ways to track and visualize web and runtime performance.

One important aspect of data gathering is deciding which format your data should be in (if you're lucky) or discovering which format your data is available in. We'll next be looking at some of the common data formats in use today.

JSON is an acronym that stands for JavaScript Object Notation. As you probably know, it is essentially a way to send data as serialized JavaScript objects. We format JSON as follows:

```
[object]{
    [attribute]: [value],
    [method] : function(){},
    [array]: [item, item]
}
```

Another way to transfer data is in XML format. XML has an expected syntax, in which elements can have attributes, which have values, values are always in quotes, and every element must have a closing element. XML looks like this:

```
<parent attribute="value">
    <child attribute="value">node data</child>
</parent>
```

Generally, we can expect APIs (or application programing interfaces) to return XML or JSON to us, and our preference is usually JSON because as we can see it is a much more lightweight option just in sheer amount of characters used.

But if we are exporting data from an application, it most likely will be in the form of a comma-separated value file, or CSV. A CSV is exactly what it sounds like: values separated by commas or some other sort of delimiter:

```
value1,value2,value3
value4,value5,value6
```

For our example, we'll use the *New York Times* API (application programming interface) tool (free registration required), available at `http://prototype.nytimes.com/gst/apitool/index.html`. The API tool exposes all the APIs that the *New York Times* makes available, including the Article Search API, the Campaign Finance API, and the Movie Review API. All we need to do is select the APIs button, then choose Article Search API button from choices presented, Choose the /articlesearch.json path, type in our search query or the phrase that we want to search for, and click "Make Request".

This queries the API and returns the data to us, formatted as JSON. We can see the results in Figure 1-16.

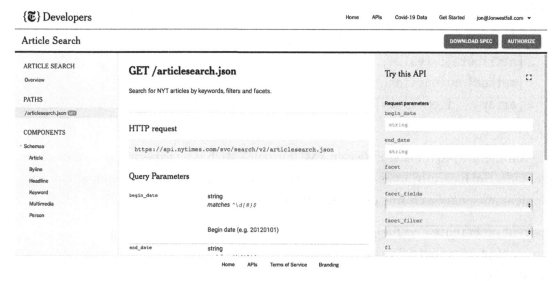

Figure 1-16. *The NY Times API tool*

We can then copy and paste the returned JSON data to our own file, or we could go the extra step to get an API key so that we can query the API from our own applications.

For the sake of our example, we will save the JSON data to a file that we will name **jsNYTimesData.txt**. The contents of the file will be structured like so:

```
{
 "offset": "0",
 "results": [
   {
     "body": "BODY COPY",
     "byline": "By AUTHOR",
     "date": "20121011",
     "title": "TITLE",
     "url": "http:\/\/ www.nytimes.com/foo.html  "
   }, {
     "body": "BODY COPY",
     "byline": "By AUTHOR",
     "date": "20121021",
     "title": "TITLE",
     "url": "http:\/\/ www.nytimes.com/bar.html  "
   }
     ],
 "tokens": [
   "JavaScript"
 ],
 "total": 2
}
```

Looking at the high-level JSON structure, we see an attribute named offset, an array named results, an array named tokens, and another attribute named total. The offset variable is for pagination (what page full of results we are starting with). The total variable is just what it sounds like: the number of results that are returned for our query. It's the results array that we really care about; it is an array of objects, each of which corresponds to an article.

The article objects have attributes named body, byline, date, title, and url.

We now have data that we can begin to look at. That takes us to our next step in the process, analyzing our data.

DATA SCRUBBING

There is often a hidden step here, one that anyone who's dealt with data knows about: scrubbing the data. Often the data is either not formatted exactly as we need it or, in even worse cases, it is dirty or incomplete.

In the best-case scenario in which your data just needs to be reformatted or even concatenated, go ahead and do that, but be sure to not lose the integrity of the data.

Dirty data has fields out of order, fields with obviously bad information in them—think dashes in ZIP codes—or gaps in the data. If your data is dirty, you have several choices:

- You could drop the rows in question, but that can harm the integrity of the data—a good example is if you are creating a histogram, removing rows could change the distribution and change what your results will be.

- The better alternative is to reach out to whoever administers the source of your data and try and get a better version if it exists.

Whatever the case, if data is dirty or it just needs to be reformatted to be able to be imported into R, expect to have to scrub your data at some point before you begin your analysis.

Analyze Data

Having data is great, but what does it mean? We determine it through analysis.

Analysis is the most crucial piece of creating data visualizations. It's only through analysis that we can understand our data, and it is only through understanding it that we can craft our story to share with others.

To begin analysis, let's import our data into R. Don't worry if you aren't completely fluent in R; we do a deep dive into the language in the next chapter. If you aren't familiar with R yet, don't worry about coding along with the following examples: just follow along to get an idea of what is happening and return to these examples after reading Chapters 3 and 4.

Because our data is JSON, let's use an R package called `rjson`. This will allow us to read in and parse JSON with the `fromJSON()` function:

```
install.packages("rjson")
library(rjson)
json_data <- fromJSON(paste(readLines("jsNYTimesData.txt"), collapse=""))
```

This is great, except the data is read in as pure text, including the date information. We can't extract information from text because obviously text has no contextual meaning outside of being raw characters. So we need to iterate through the data and parse it to more meaningful types.

Let's create a data frame (an array-like data type specific to R that we talk about next chapter), loop through our `json_data` object, and parse year, month, and day parts out of the `date` attribute. Let's also parse the author name out of the `byline` and check to make sure that if the author's name isn't present, we substitute the empty value with the string `"unknown"`.

```
df <- data.frame()
for(n in json_data$response$docs){
    year <-substr(n$pub_date, 0, 4)
    month <- substr(n$pub_date, 6, 7)
    day <- substr(n$pub_date, 9, 10)
    author <- substr(n$byline$original, 4, 30)
    title <- n$headline$main
    if(length(author) < 1){
            author <- "unknown"
    }
```

Next, we can reassemble the date into a *MM/DD/YYYY* formatted string and convert it to a date object:

```
datestamp <-paste(month, "/", day, "/", year, sep="")
datestamp <- as.Date(datestamp,"%m/%d/%Y")
```

And finally, before we leave the loop, we should add this newly parsed author and date information to a temporary row and add that row to our new data frame:

```
      newrow <- data.frame(datestamp, author, title,
      stringsAsFactors=FALSE, check.rows=FALSE)
      df <- rbind(df, newrow)
}
rownames(df) <- df$datestamp
Our complete loop should look like the following:
df <- data.frame()
for(n in json_data$response$docs){
      year <-substr(n$pub_date, 0, 4)
      month <- substr(n$pub_date, 6, 7)
      day <- substr(n$pub_date, 9, 10)
      author <- substr(n$byline$original, 4, 30)
      title <- n$headline$main
      if(length(author) < 1){
            author <- "unknown"
      }
      datestamp <-paste(month, "/", day, "/", year, sep="")
      datestamp <- as.Date(datestamp,"%m/%d/%Y")
      newrow <- data.frame(datestamp, author, title,
      stringsAsFactors=FALSE, check.rows=FALSE)
      df <- rbind(df, newrow)
}
rownames(df) <- df$datestamp
```

Note that our example assumes that the data set returned has unique date values. If you get errors with this, you may need to scrub your returned data set to purge any duplicate rows. Also be mindful that the *New York Times* API may change over time. Between revisions of this book, the API tool changed various titles (e.g., "title" became "headline"). If this code doesn't appear to work, you'll want to read through the JSON data to see if, perhaps, they've pulled a switch again!

Once our data frame is populated, we can start to do some analysis on the data. Let's start out by pulling just the year from every entry and quickly making a stem and leaf plot to see the shape of the data.

Note   John Tukey created the stem and leaf plot in his seminal work, *Exploratory Data Analysis*. Stem and leaf plots are quick, high-level ways to see the shape of data, much like a histogram. In the stem and leaf plot, we construct the "stem" column on the left and the "leaf" column on the right. The stem consists of the most significant unique elements in a result set. The leaf consists of the remainder of the values associated with each stem. In our stem and leaf plot in the following, the years are our stem and R shows zeroes for each row associated with a given year. Something else to note is that often alternating sequential rows are combined into a single row, in the interest of having a more concise visualization.

First, we will create a new variable to hold the year information:

```
yearlist <- as.POSIXlt(df$datestamp)$year+1900
```

If we inspect this variable, we see that it looks something like this:

```
> yearlist
  [1] 2012 2012 2012 2012 2012 2012 2012 2012 2012 2012 2012 2012 2012 2011
 2011 2011 2011 2011 2011 2011 2011 2011 2011 2011 2011 2011 2011 2011 2011
 [30] 2011 2011 2011 2011 2010 2010 2010 2010 2010 2010 2010 2010 2010 2010
 2009 2009 2009 2009 2009 2009 2009 2008 2008 2008 2007 2007 2007 2007 2006
 [59] 2006 2006 2006 2005 2005 2005 2005 2005 2005 2004 2003 2003 2003 2002
 2002 2002 2002 2001 2001 2000 2000 2000 2000 2000 2000 1999 1999 1999 1999
 [88] 1999 1999 1998 1998 1998 1997 1997 1996 1996 1995 1995 1995 1993 1993
 1993 1993 1992 1991 1991 1991 1990 1990 1990 1990 1989 1989 1989 1988 1988
 [117] 1988 1986 1985 1985 1985 1984 1982 1982 1981
```

That's great, that's exactly what we want: a year to represent every article returned. Next, let's create the stem and leaf plot:

```
> stem(yearlist)
  1980 | 0
  1982 | 00
  1984 | 0000
  1986 | 0
  1988 | 000000
```

```
1990 | 0000000
1992 | 00000
1994 | 000
1996 | 0000
1998 | 000000000
2000 | 00000000
2002 | 0000000
2004 | 0000000
2006 | 00000000
2008 | 0000000000
2010 | 000000000000000000000000000000000
2012 | 000000000000
```

Very interesting. We see a gradual build with some dips in the mid-1990s, another gradual build with another dip in the mid-2000s, and a strong explosion since 2010 (the stem and leaf plot groups years together in twos).

Looking at that, my mind starts to envision a story building about a subject growing in popularity. But what about the authors of these articles? Maybe they are the result of one or two very interested authors that have quite a bit to say on the subject.

Let's explore that idea and take a look at the author data that we parsed out. Let's look at just the unique authors from our data frame:

```
> length(unique(df$author))
[1] 81
```

We see that there are 81 unique authors or combination of authors for these articles! Just out of curiosity, let's take a look at the breakdown by author for each article. Let's quickly create a bar chart to see the overall shape of the data (the bar chart is shown in Figure 1-17):

```
plot(table(df$author), axes=FALSE)
```

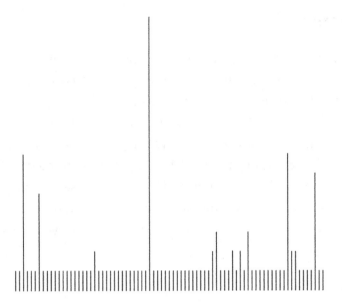

Figure 1-17. *Bar chart of number of articles by author to quickly visualize*

We remove the x- and y-axes to allow ourselves to focus just on the shape of the data without worrying too much about the granular details. From the shape, we can see a large number of bars with the same value; these are authors who have written a single article. The higher bars are authors who have written multiple articles. Essentially each bar is a unique author, and the height of the bar indicates the number of articles they have written. We can see that although there are roughly five standout contributors, most authors have average one article.

Note that we just created several visualizations as part of our analysis. The two steps aren't mutually exclusive; we oftentimes create quick visualizations to facilitate our own understanding of the data. It's the intention with which they are created that make them part of the analysis phase. These visualizations are intended to improve our own understanding of the data so that we can accurately tell the story in the data.

What we've seen in this particular data set tells a story of a subject growing in popularity, demonstrated by the increasing number of articles (in the stem plot) by a variety of authors (in the bar plot). Let's now prepare it for mass consumption.

Note   We are not fabricating or inventing this story. Like information archaeologists, we are sifting through the raw data to uncover the story.

Visualize Data

Once we've analyzed the data and understand it (and I mean really understand the data to the point where we are conversant in all the granular details around it), and once we've seen the story that the data has within, it is time to share that story.

For the current example, we've already crafted a stem and leaf plot as well as a bar chart as part of our analysis. However, stem and leaf plots are great for analyzing data, but not so great for messaging out about the findings. It is not immediately obvious what the context of the numbers in a stem and leaf plot represents. And the bar chart we created supported the main thesis of the story instead of communicating that thesis.

Since we want to demonstrate the distribution of articles by year, let's instead use a histogram to tell the story:

```
hist(yearlist)
```

See Figure 1-18 for what this call to the hist() function generates.

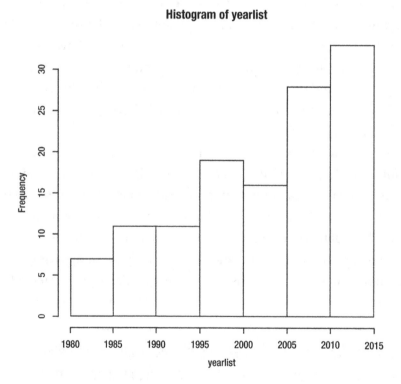

Figure 1-18. *Histogram of yearlist*

This is a good start, but let's refine this further. Let's color in the bars, give the chart a meaningful title, and strictly define the range of years:

```
hist(yearlist, breaks=(1981:2012), freq=TRUE, col="#CCCCCC",
main="Distribution of Articles about Data Visualization\nby the NY Times",
xlab = "Year")
```

This produces the histogram that we see in Figure 1-5.

Ethics of Data Visualization

Remember Figure 1-3 from the beginning of this chapter where we looked at the weighted popularity of the search term "Data Visualization"? By constraining the data to 2006 to 2012, we told a story of a keyword growing in popularity, almost doubling in popularity over a six-year period. But what if we included more data points in our sample and extended our view to include 2004? See Figure 1-19 for this expanded time series chart.

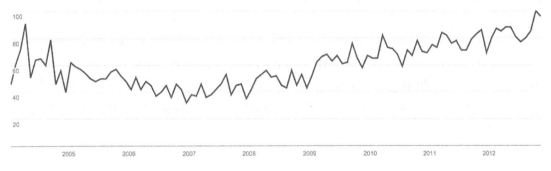

Figure 1-19. *Google Trends time series chart with expanded time range. Note that the additional data points give a greater context and tell a different story*

This expanded chart tells a different story: one that describes a dip in popularity between 2005 and 2009. This expanded chart also demonstrates how easy it is to misrepresent the truth intentionally or unintentionally with data visualizations.

Cite Sources

When Playfair first published his *Commercial and Political Atlas,* one of the biggest biases he had to battle was the inherent distrust his peers had of charts to accurately represent data. He tried to overcome this by including data tables in the first two editions of the book.

Similarly, we should always include our sources when distributing our charts so that our audience can go back and independently verify the data if they want to. This is important because we are trying to share information, not hoard it, and we should encourage others to inspect the data for themselves and be excited about the results.

Be Aware of Visual Cues

A side effect of using charts to function as visual shorthand is that we bring our own perspective and context to play when we view charts. We are used to certain things, such as the color red being used to signify danger or flagging for attention or the color green signifying safety. These color connotations are part of a branch of color theory called color harmony, and it's worth at least being aware of what your color choices could be implying.

When in doubt, get a second opinion. When creating our graphics, we can often get married to a certain layout or chart choice. This is natural because we have spent time invested in analyzing and crafting the chart. A fresh, objective set of eyes should point out unintentional meanings or overly complex designs and make for a more crisp visualization.

Summary

This chapter took a look at some introductory concepts about data visualization, from conducting data gathering and exploration to looking at the charts that make up the visual patterns that define how we communicate with data. We looked a little at the history of data visualization, from the early beginnings with William Playfair and Florence Nightingale to modern examples such as xkcd.com.

While we saw a little bit of code in this chapter, in the next chapter we start to dig in to the tactics of learning R and getting our hands dirty reading in data, shaping data, and crafting our own visualizations.

CHAPTER 2

R Language Primer

In the last chapter, we defined what data visualizations are, looked at a little bit of the history of the medium, and explored the process for creating them. This chapter takes a deeper dive into one of the most important tools for creating data visualizations: R.

When creating data visualizations, R is an integral tool for both analyzing data and creating visualizations. We will use R extensively through the rest of this book, so we had better level set first.

R is both an environment and a language to run statistical computations and produce data graphics. It was created by Ross Ihaka and Robert Gentleman in 1993 while at the University of Auckland. The R environment is the runtime environment that you develop and run R in. The R language is the programming language that you develop in.

R is the successor to the S language, a statistical programming language that came out of Bell Labs in 1976.

Getting to Know the R Console

Let's start by downloading and installing R. R is available from the R Foundation at `www.r-project.org/`. See Figure 2-1 for a screenshot of the R Foundation home page.

© Tom Barker, Jon Westfall 2022
T. Barker and J. Westfall, *Pro Data Visualization Using R and JavaScript*,
https://doi.org/10.1007/978-1-4842-7202-2_2

The R Project for Statistical Computing

[Home]

Download

CRAN

R Project

About R
Logo
Contributors
What's New?
Reporting Bugs
Conferences
Search
Get Involved: Mailing Lists
Developer Pages
R Blog

R Foundation

Foundation
Board
Members
Donors
Donate

Help With R

Getting Help

Getting Started

R is a free software environment for statistical computing and graphics. It compiles and runs on a wide variety of UNIX platforms, Windows and MacOS. To **download R**, please choose your preferred CRAN mirror.

If you have questions about R like how to download and install the software, or what the license terms are, please read our answers to frequently asked questions before you send an email.

News

- **R version 4.1.0 (Camp Pontanezen) prerelease versions** will appear starting Saturday 2021-04-17. Final release is scheduled for Tuesday 2021-05-18.
- **R version 4.0.5 (Shake and Throw)** has been released on 2021-03-31.
- Thanks to the organisers of useR! 2020 for a successful online conference. Recorded tutorials and talks from the conference are available on the R Consortium YouTube channel.
- **R version 3.6.3 (Holding the Windsock)** was released on 2020-02-29.
- You can support the R Foundation with a renewable subscription as a supporting member

News via Twitter

News from the R Foundation

Figure 2-1. *Home page of the R Foundation*

It is available as a precompiled binary from the Comprehensive R Archive Network (CRAN) website: http://cran.r-project.org/ (see Figure 2-2). We just select our operating system and what version of R we want, and we can begin to download.

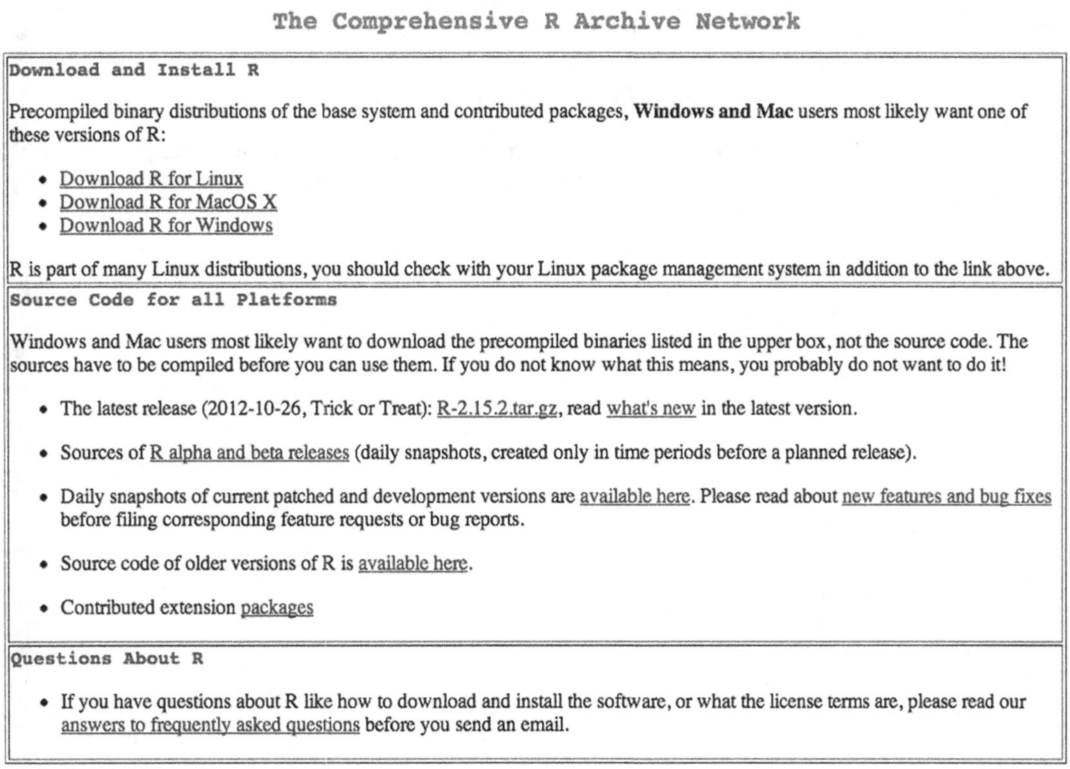

The Comprehensive R Archive Network

Download and Install R
Precompiled binary distributions of the base system and contributed packages, **Windows and Mac** users most likely want one of these versions of R: • Download R for Linux   • Download R for MacOS X   • Download R for Windows   R is part of many Linux distributions, you should check with your Linux package management system in addition to the link above.
Source Code for all Platforms
Windows and Mac users most likely want to download the precompiled binaries listed in the upper box, not the source code. The sources have to be compiled before you can use them. If you do not know what this means, you probably do not want to do it!   • The latest release (2012-10-26, Trick or Treat): R-2.15.2.tar.gz, read what's new in the latest version.   • Sources of R alpha and beta releases (daily snapshots, created only in time periods before a planned release).   • Daily snapshots of current patched and development versions are available here. Please read about new features and bug fixes before filing corresponding feature requests or bug reports.   • Source code of older versions of R is available here.   • Contributed extension packages
Questions About R
• If you have questions about R like how to download and install the software, or what the license terms are, please read our answers to frequently asked questions before you send an email.

Figure 2-2. *The CRAN website*

Once the download is complete, we can run through the installer. See Figure 2-3 for a screenshot of the R installer for macOS.

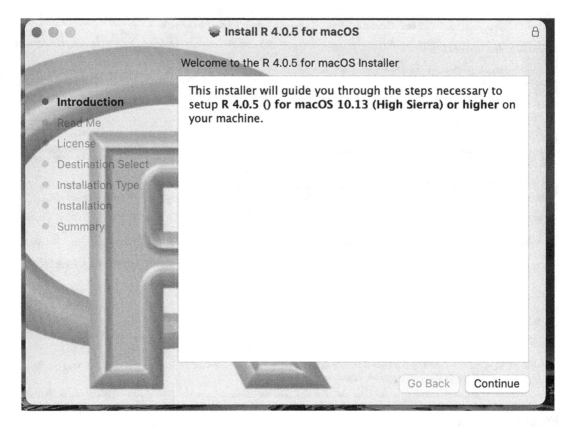

Figure 2-3. *R installation on a Mac*

Once we finish the installation, we can launch the R application, and we are presented with the R console, as shown in Figure 2-4.

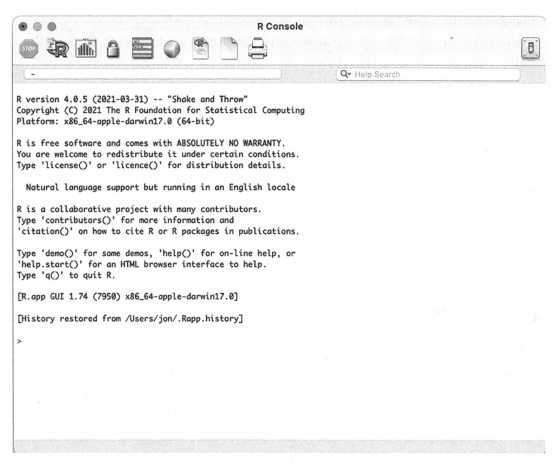

Figure 2-4. *The R console*

The Command Line

The R console is where the magic happens! It is a command-line environment where we can run R expressions. The best way to get up to speed in R is to script in the console, a piece at a time, generally to try out what you're trying to do, and tweak it until you get the results that you want. When you finally have a working example, take the code that does what you want and save it as an R script file.

R script files are just files that contain pure R and can be run in the console using the source command:

```
> source("someRfile.R")
```

Looking at the preceding code snippet, we assume that the R script lives in the current work directory. The way we can see what the current work directory is to use the getwd() function:

```
> getwd()
[1] "/Users/tomjbarker"
```

We can also set the working directory by using the setwd() function. Note that changes made to the working directory are not persisted across R sessions unless the session is saved.

```
> setwd("/Users/tomjbarker/Downloads")
> getwd()
[1] "/Users/tomjbarker/Downloads"
```

Command History

The R console stores commands that you enter and you can cycle through previous commands by pressing the up arrow. Hit the escape button to return to the command prompt. We can see the history in a separate window pane by clicking the Show/Hide Command History button at the top of the console. The Show/Hide Command History button is the rectangle icon with alternating stripes of yellow and green. See Figure 2-5 for the R console with the command history shown.

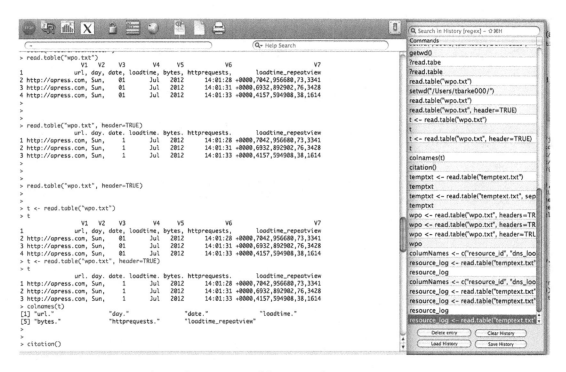

Figure 2-5. *R console with command history shown*

Accessing Documentation

To read the R documentation around a specific function or keyword, you simply type a question mark before the keyword:

```
> ?setwd
```

If you want to search the documentation for a specific word or phrase, you can type two question marks before the search query:

```
> ??"working directory"
```

This code launches a window that shows search results (see Figure 2-6). The search result window has a row for each topic that contains the search phrase and has the name of the help topic, the package that the functionality that the help topic talks about is in, and a short description for the help topic.

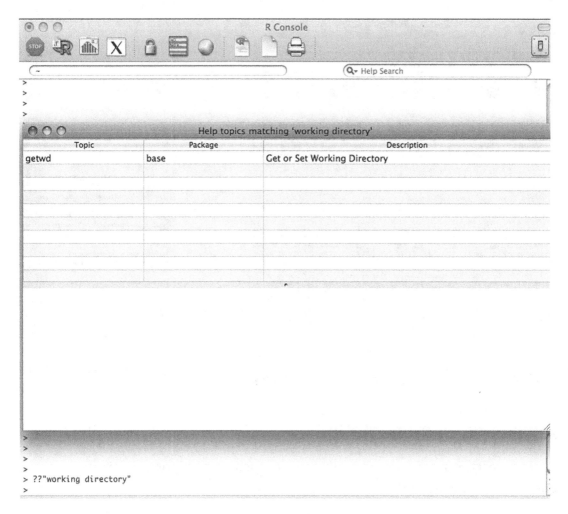

Figure 2-6. *Help search results window*

Packages

Speaking of packages, what are they, exactly? *Packages* are collections of functions, data sets, or objects that can be imported into the current session or workspace to extend what we can do in R. Anyone can make a package and distribute it.

To install a package, we simply type this:

```
install.packages([package name])
```

For example, if we want to install the `ggplot2` package—which is a widely used and very handy charting package—we simply type this into the console:

```
> install.packages("ggplot2")
```

We are immediately prompted to choose the mirror location that we want to use, usually the one closest to our current location. From there, the install begins. We can see the results in Figure 2-7.

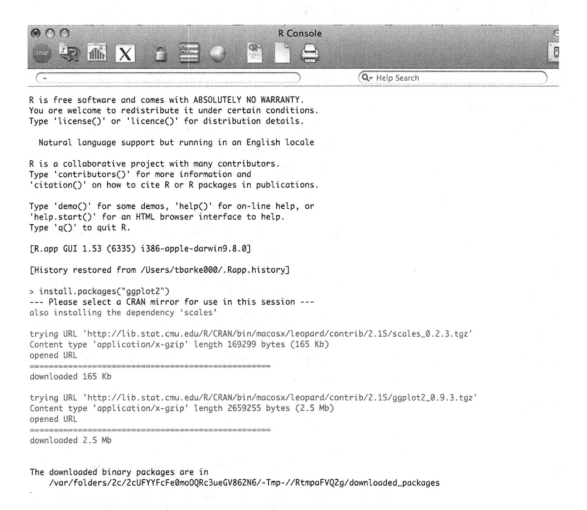

Figure 2-7. *Installing the ggplot2 package*

The zipped up package is downloaded and exploded into our R installation.

If we want to use a package that we have installed, we must first include it in our workspace. To do this, we use the library() function:

```
> library(ggplot2)
```

A list of packages available at the CRAN can be found here: http://cran.r-project.org/web/packages/available_packages_by_name.html.

To see a list of packages already installed, we can simply call the library() function with no parameter (depending on your install and your environment, your list of packages may vary):

```
> library()
Packages in library '/Library/Frameworks/R.framework/Versions/2.15/
Resources/library':
barcode                    Barcode distribution plots
base                       The R Base Package
boot                       Bootstrap Functions (originally by Angelo
                           Canty for S)
class                      Functions for Classification
cluster                    Cluster Analysis Extended Rousseeuw et al.
codetools                  Code Analysis Tools for R
colorspace                 Color Space Manipulation
compiler                   The R Compiler Package
datasets                   The R Datasets Package
dichromat                  Color schemes for dichromats
digest                     Create cryptographic hash digests of R
                           objects
foreign                    Read Data Stored by Minitab, S, SAS, SPSS,
                           Stata, Systat, dBase,
                           ...
ggplot2                    An implementation of the Grammar of
                           Graphics
gpairs                     gpairs: The Generalized Pairs Plot
graphics                   The R Graphics Package
grDevices                  The R Graphics Devices and Support for
                           Colours and Fonts
grid                       The Grid Graphics Package
```

gtable	Arrange grobs in tables
KernSmooth	Functions for kernel smoothing for Wand & Jones (1995)
labeling	Axis Labeling
lattice	Lattice Graphics
mapdata	Extra Map Databases
mapproj	Map Projections
maps	Draw Geographical Maps

Importing Data

So now our environment is downloaded and installed, and we know how to install any packages that we may need. Now we can begin using R.

The first thing we'll normally want to do is import your data. There are several ways to import data, but the most common way is to use the read() function, which has several flavors:

```
read.table("[file to read]")
read.csv(["file to read"])
```

To see this in action, let's first create a text file named temptext.txt that is formatted like so:

```
134,432,435,313,11
403,200,500,404,33
77,321,90,2002,395
```

We can read this into a variable that we will name temptxt:

```
> temptxt <- read.table("temptext.txt")
```

Notice that as we are assigning value to this variable, we are not using an equal sign as the assignment operator. We are instead using an arrow <-. That is R's assignment operator, although it does also support the equal sign if you are so inclined. But the standard is the arrow, and all examples that we will show in this book will use the arrow.

If we print out the `temptxt` variable, we see that it is structured as follows:

```
> temptxt
                    V1
1 134,432,435,313,11
2 403,200,500,404,33
3 77,321,90,2002,395
```

We see that our variable is a table-like structure called a *data frame*, and R has assigned a column name (V1) and row IDs to our data structure. More on column names soon.

The `read()` function has a number of parameters that you can use to refine how the data is imported and formatted once it is imported.

Using Headers

The `header` parameter tells R to treat the first line in the external file as containing header information. The first line then becomes the column names of the data frame.

For example, suppose we have a log file structured like this:

```
url, day, date, loadtime, bytes, httprequests, loadtime_repeatview
http://apress.com , Sun, 01 Jul 2012 14:01:28 +0000,7042,956680,73,3341
http://apress.com , Sun, 01 Jul 2012 14:01:31 +0000,6932,892902,76,3428
http://apress.com , Sun, 01 Jul 2012 14:01:33 +0000,4157,594908,38,1614
```

We can load it into a variable named wpo like so:

```
> wpo <- read.table("wpo.txt", header=TRUE, sep=",")
> wpo
  url  day date loadtime bytes httprequests loadtime_repeatview
```

1. http://apress.com,Sun,1 Jul 2012 14:01:28
   +0000,7042,955550,73,3191

2. http://apress.com,Sun,1 Jul 2012 14:01:31
   +0000,6932,892442,76,3728

3. http://apress.com,Sun,1 Jul 2012 14:01:33
   +0000,4157,614908,38,1514

When we call the `colnames()` function to see what the column names are for `wpo`, we see the following:

```
> colnames(wpo)
[1] "url"       "day"          "date"         "loadtime"
[5] "bytes"       "httprequests"       "loadtime_repeatview"
```

Specifying a String Delimiter

The `sep` attribute tells the `read()` function what to use as the string delimiter for parsing the columns in the external data file. In all the examples we've looked at so far, commas are our delimiters (as we explicitly told R in the line that read in wpo), but we could use instead pipes | or any other character that we want.

Say, for example, that our previous `temptxt` example used pipes; we would just update the code to be as follows:

```
134|432|435|313|11
403|200|500|404|33
77|321|90|2002|395
> temptxt <- read.table("temptext.txt", sep="|")
> temptxt
  V1  V2  V3   V4  V5
```

1. 134 432 435 313 11

2. 403 200 500 404 33

3. 77 321 90 2002 395

Oh, notice that? We actually got distinct column names this time (V1, V2, V3, V4, V5). Before, we didn't specify a delimiter, so R assumed that each row was one big blob of text and lumped it into a single column (V1).

Specifying Row Identifiers

The `row.names` attribute allows us to specify identifiers for our rows. By default, as we've seen in the previous examples, R uses incrementing numbers as row IDs. Keep in mind that the row names need to be unique for each row.

With that in mind, let's take a look at importing some different log data, which has performance metrics for unique URLs:

```
url, day, date, loadtime, bytes, httprequests, loadtime_repeatview
http://apress.com, Sun, 01 Jul 2012 14:01:28 +0000,7042,956680,73,3341
http://google.com, Sun, 01 Jul 2012 14:01:31 +0000,6932,892902,76,3428
http://apple.com, Sun, 01 Jul 2012 14:01:33 +0000,4157,594908,38,1614
```

When we read it in, we'll be sure to specify that the data in the url column should be used as the row name for the data frame:

```
> wpo <- read.table("wpo.txt", header=TRUE, sep=",", row.names="url")
> wpo
                   day  date                          loadtime  bytes
                        httprequests  loadtime_repeatview
http://apress.com  Sun  01 Jul 2012 14:01:28  +0000  7042      956680
                        73            3341
http://google.com  Sun  01 Jul 2012 14:01:31  +0000  6932      892902
                        76            3428
http://apple.com   Sun  01 Jul 2012 14:01:33  +0000  4157      594908
                        38            31614
```

Using Custom Column Names

And there we go. But what if we want to have column names, but the first line in our file is not header information? We can use the col.names parameter to specify a vector that we can use as column names.

Let's take a look. In this example, we'll use the pipe-separated text file used previously:

```
134|432|435|313|11
403|200|500|404|33
77|321|90|2002|395
```

First, we'll create a vector named columnNames that will hold the strings that we will use as the column names:

```
> columnNames <- c("resource_id", "dns_lookup", "cache_load", "file_size",
"server_response")
```

Then, we'll read in the data, passing in our vector to the col.names parameter:

```
> resource_log <- read.table("temptext.txt", sep="|", col.
names=columnNames)
> resource_log
  resource_id dns_lookup cache_load file_size server_response
1         134        432        435       313              11
2         403        200        500       404              33
3          77        321         90      2002             395
```

Data Structures and Data Types

In the previous examples, we touched on a lot of concepts; we created variables, including vectors and data frames; but we didn't talk much about what they are. Let's take a step back and look at the data types that R supports and how to use them.

Data types in R are called *modes* and can be the following:

- Numeric

- Character

- Logical

- Complex

- Raw

- List

We can use the mode() function to check the mode of a variable.

Character and numeric modes correspond to string and number (both integer and float) data types. Logical modes are Boolean values.

```
> n <- 122132
> mode(n)
[1] "numeric"
> c <- "test text"
> mode(c)
[1] "character"
> l <- TRUE
> mode(l)
[1] "logical"
```

We can perform string concatenation using the paste() function. We can use the substr() function to pull characters out of strings. Let's look at some examples in code.

Usually, I keep a list of directories that I either read data from or write charts to. Then when I want to reference a new data file that exists in the data directory, I will just append the new file name to the data directory:

```
> dataDirectory <- "/Users/tomjbarker/org/data/"
> buglist <- paste(dataDirectory, "bugs.txt", sep="")
> buglist
[1] "/Users/tomjbarker/org/data/bugs.txt"
```

The paste() function takes N amount of strings and concatenates them together. It accepts an argument named sep that allows us to specify a string that we can use to be a delimiter between joined strings. We don't want anything separating our joined strings that we pass in an empty string.

If we want to pull characters from a string, we use the substr() function. The substr() function takes a string to parse, a starting location, and a stopping location. It returns all the character inclusively from the starting location up to the ending location. (Remember that in R, lists are not 0 based like most other languages, but instead have a starting index of 1.)

```
> substr("test", 1,2)
[1] "te"
```

In the preceding example, we pass in the string "test" and tell the substr() function to return the first and second characters.

Complex mode is for complex numbers. The raw mode is to store raw byte data.

List data types or modes can be one of three classes: vectors, matrices, or data frames. If we call mode() for vectors or matrices, they return the mode of the data that they contain; class() returns the class. If we call mode() on a data frame, it returns the type list.

```
> v <- c(1:10)
> mode(v)
[1] "numeric"
> m <- matrix(c(1:10), byrow=TRUE)
> mode(m)
[1] "numeric"
> class(m)
[1] "matrix" "array"
> d <- data.frame(c(1:10))
> mode(d)
[1] "list"
> class(d)
[1] "data.frame"
```

Note that we just typed 1:10 rather than the whole sequence of numbers between 1 and 10:

```
v <- c(1:10)
```

Vectors are single-dimensional arrays that can hold only values of a single mode at a time. It's when we get to data frames and matrices that R really starts to get interesting. The next two sections cover those classes.

Data Frames

We saw at the beginning of this chapter that the read() function takes in external data and saves it as a data frame. *Data frames* are like arrays in most other loosely typed languages: they are containers that hold different types of data, referenced by index. The main thing to realize, though, is that data frames see the data that they contain as rows, columns, and combinations of the two.

For example, think of a data frame as formatted as follows:

```
      col  col  col  col  col
row [ 1 ] [ 1 ] [ 1 ] [ 1 ] [ 1 ]
row [ 1 ] [ 1 ] [ 1 ] [ 1 ] [ 1 ]
row [ 1 ] [ 1 ] [ 1 ] [ 1 ] [ 1 ]
row [ 1 ] [ 1 ] [ 1 ] [ 1 ] [ 1 ]
```

If we try to reference the first index in the preceding data frame as we traditionally would with an array, say dataframe[1], R would instead return the first column of data, not the first item. So data frames are referenced by their column and row. So dataframe[1] returns the first column, and dataframe[,2] returns the first row.

Let's demonstrate this in code.

First, let's create some vectors using the combine function, c(). Remember that vectors are collections of data all of the same type. The combine function takes a series of values and combines them into vectors.

```
> col1 <- c(1,2,3,4,5,6,7,8)
> col2 <- c(1,2,3,4,5,6,7,8)
> col3 <- c(1,2,3,4,5,6,7,8)
> col4 <- c(1,2,3,4,5,6,7,8)
```

Then, let's combine these vectors into a data frame:

```
> df <- data.frame(col1,col2,col3,col4)
```

Now let's print the data frame to see the contents and the structure of it:

```
> df
  col1 col2 col3 col4
1   1    1    1    1
2   2    2    2    2
3   3    3    3    3
4   4    4    4    4
5   5    5    5    5
6   6    6    6    6
7   7    7    7    7
8   8    8    8    8
```

Notice that it took each vector and made each one a column. Also notice that each row has an ID; by default, it is a number, but we can override that.

If we reference the first index, we see that the data frame returns the first column:

```
> df[1]
  col1
1    1
2    2
3    3
4    4
5    5
6    6
7    7
8    8
```

If we put a comma in front of that 1, we reference the first row:

```
> df[,1]
[1] 1 2 3 4 5 6 7 8
```

So accessing contents of a data frame is done by specifying [column, row]. Matrices work much the same way.

Matrices

Matrices are just like data frames in that they contain rows and columns and can be referenced by either. The core difference between the two is that data frames can hold different data types, but matrices can hold only one type of data.

This presents a philosophical difference. Usually, you use data frames to hold data read in externally, like from a flat file or a database because those are generally of mixed type. You normally store data in matrices that you want to apply functions to (more on applying functions to lists in a little bit).

To create a matrix, we must use the matrix() function, pass in a vector, and tell the function how to distribute the vector:

- The nrow parameter specifies how many rows the matrix should have.

- The ncol parameter specifies the number of columns.

- The byrow parameter tells R that the contents of the vector should be distributed by iterating across rows if TRUE or by columns if FALSE.

```
> content <- c(1,2,3,4,5,6,7,8,9,10)
> m1 <- matrix(content, nrow=2, ncol=5, byrow=TRUE)
> m1
     [,1] [,2] [,3] [,4] [,5]
[1,]    1    2    3    4    5
[2,]    6    7    8    9   10
>
```

Notice that in the previous example the m1 matrix is filled in horizontally, row by row. In the following example, the m1 matrix is filled in vertically by column:

```
> content <- c(1,2,3,4,5,6,7,8,9,10)
> m1 <- matrix(content, nrow=2, ncol=5, byrow=FALSE)
> m1
     [,1] [,2] [,3] [,4] [,5]
[1,]    1    3    5    7    9
[2,]    2    4    6    8   10
```

Remember that instead of manually typing out all the numbers in the previous content vector, if the numbers are a sequence, we can just type this:

```
content <- (1:10)
```

We reference the content in matrices with the square bracket, specifying the row and column, respectively:

```
> m1[1,4]
[1] 7
```

We can convert a data frame to a matrix if the data frame contains only a single type of data. To do this, we use the `as.matrix()` function. Oftentimes, we will do this when passing a data frame to a plotting function to draw a chart.

```
> barplot(as.matrix(df))
```

In the following, we create a data frame called df. We populate the data frame with ten consecutive numbers. We then use `as.matrix()` to convert df into a matrix and save the result into a new variable called m.

```
> df <- data.frame(1:10)
> df
  X1.10
1      1
2      2
3      3
4      4
5      5
6      6
7      7
8      8
9      9
10    10
> class(df)
[1] "data.frame"
> m <- as.matrix(df)
> class(m)
[1] "matrix" "array"
```

Keep in mind that because they are all the same data type, matrices require less overhead and are intrinsically more efficient than data frames. If we compare the size of our matrix m and our data frame df, we see that with just ten items, there is a size difference.

```
> object.size(m)
552 bytes
> object.size(df)
776 bytes
```

With that said, if we increase the scale of this, the increase in efficiency does not equally scale. Compare the following:

```
> big_df <- data.frame(1:40000000)
> big_m <- matrix(1:40000000)
> object.size(big_m)
160000216 bytes
> object.size(big_df)
160000736 bytes
```

We can see that the first example with the small data set showed that the matrix was 30 percent smaller in size than the data frame, but at the larger scale in the second example, the matrix was only .00018 percent smaller than the data frame.

Adding Lists

When combining or adding to data frames or matrices, you generally add either by the row or the column using rbind() or cbind().

To demonstrate this, let's add a new row to our data frame df. We'll pass df into rbind() along with the new row to add to df. The new row contains just one element, the number 11.

```
> df <- rbind(df, 11)
> df
  X1.10
1      1
2      2
3      3
4      4
5      5
6      6
7      7
8      8
9      9
10    10
11    11
```

Now let's add a new column to our matrix m. To do this, we simply pass m into cbind() as the first parameter; the second parameter is a new matrix that will be appended to the new column.

```
> m <- rbind(m, 11)
> m <- cbind(m, matrix(c(50:60), byrow=FALSE))
> m
      X1.10
[1,]      1  50
[2,]      2  51
[3,]      3  52
[4,]      4  53
[5,]      5  54
[6,]      6  55
[7,]      7  56
[8,]      8  57
[9,]      9  58
[10,]    10  59
[11,]    11  60
```

What about vectors, you may ask? Well, let's look at adding to our content vector. We simply use the combine function to combine the current vector with a new vector:

```
> content <- c(1,2,3,4,5,6,7,8,9,10)
> content <- c(content, c(11:20))
> content
[1]  1  2  3  4  5  6  7  8  9 10 11 12 13 14 15 16 17 18 19 20
```

Looping Through Lists

As developers who generally work in procedural languages, or at least came up the ranks using procedural languages (though, in recent years, functional programming paradigms have become much more mainstream), we're most likely used to looping through our arrays when we want to process the data within them. This is in contrast to purely functional languages where we would instead apply a function to our lists, like the map() function. R supports both paradigms. Let's first look at how to loop through our lists.

The most useful loop that R supports is the for in loop. The basic structure of a for in loop can be seen here:

```
> for(i in 1:5){print(i)}
[1] 1
[1] 2
[1] 3
[1] 4
[1] 5
```

The variable i increments in value each step through the iteration. We can use the for in loop to step through lists. We can specify a particular column to iterate through, like the following, in which we loop through the X1.10 column of the data frame df.

```
> for(n in df$X1.10){ print(n)}
[1] 1
[1] 2
[1] 3
[1] 4
[1] 5
[1] 6
[1] 7
[1] 8
[1] 9
[1] 10
[1] 11
```

Note that we are accessing the columns of data frames via the dollar sign operator. The general pattern is [data frame]$[column name].

Applying Functions to Lists

But the way that R really wants to be used is to apply functions to the contents of lists (see Figure 2-8).

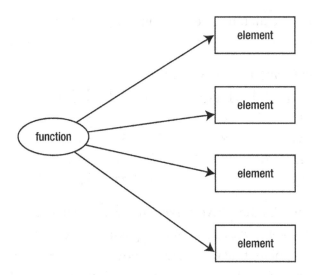

Figure 2-8. *Apply a function to list elements*

We do this in R with the apply() function.

The apply() function takes several parameters:

- First is our list.

- Next, a number vector to indicate how we apply the function through the list (1 is for rows, 2 is for columns, and c[1,2] indicates both rows and columns).

- Last is the function to apply to the list:

```
apply([list], [how to apply function], [function to apply])
```

Let's look at an example. Let's make a new matrix that we'll call m. The matrix m will have ten columns and four rows:

```
> m <- matrix(c(1:40), byrow=FALSE, ncol=10)
> m
     [,1] [,2] [,3] [,4] [,5] [,6] [,7] [,8] [,9] [,10]
[1,]    1    5    9   13   17   21   25   29   33    37
[2,]    2    6   10   14   18   22   26   30   34    38
[3,]    3    7   11   15   19   23   27   31   35    39
[4,]    4    8   12   16   20   24   28   32   36    40
```

Now say we wanted to increment every number in the m matrix. We could simply use apply() as follows:

```
> apply(m, 2, function(x) x <- x + 1)
     [,1] [,2] [,3] [,4] [,5] [,6] [,7] [,8] [,9] [,10]
[1,]    2    6   10   14   18   22   26   30   34    38
[2,]    3    7   11   15   19   23   27   31   35    39
[3,]    4    8   12   16   20   24   28   32   36    40
[4,]    5    9   13   17   21   25   29   33   37    41
```

Do you see what we did there? We passed in m, we specified that we wanted to apply the function across the columns, and finally we passed in an *anonymous function*. The function accepts a parameter that we called x. The parameter x is a reference to the current matrix element. From there, we just increment the value of x by 1.

OK, say we wanted to do something slightly more interesting, such as zeroing out all the even numbers in the matrix. We could do the following:

```
> apply(m,c(1,2),function(x){if((x %% 2) == 0) x <- 0 else x <- x})
     [,1] [,2] [,3] [,4] [,5] [,6] [,7] [,8] [,9] [,10]
[1,]    1    5    9   13   17   21   25   29   33    37
[2,]    0    0    0    0    0    0    0    0    0     0
[3,]    3    7   11   15   19   23   27   31   35    39
[4,]    0    0    0    0    0    0    0    0    0     0
```

For the sake of clarity, let's break out that function that we are applying. We simply check to see whether the current element is even by checking to see whether it has a remainder when divided by two. If it is, we set it to zero; if it isn't, we set it to itself:

```
function(x){
    if((x %% 2) == 0)
        x <- 0
    else
        x <- x
}
```

Functions

Speaking of functions, the syntax for creating functions in R is much like most other languages. We use the `function` keyword, give the function a name, have open and closed parentheses where we specify arguments, and wrap the body of the function in curly braces:

```
function [function name]([argument])
{
    [body of function]
}
```

Something interesting that R allows is the ... argument (sometimes called the dots argument). This allows us to pass in a variable number of parameters into a function. Within the function, we can convert the ... argument into a list and iterate over the list to retrieve the values within:

```
> offset <- function (...){
    for(i in list(...)){
        print(i)
    }
}
> offset(23,11)
[1] 23
[1] 11
```

We can even store values of different data types (modes) in the ... argument:

```
> offset("test value", 12, 100, "19ANM")
[1] "test value"
[1] 12
[1] 100
[1] "19ANM"
```

R uses lexical scoping. This means that when we call a function and try to reference variables that are not defined inside the local scope of the function, the R interpreter looks for those variables in the workspace or scope in which the function was created. If the R interpreter cannot find those variables in that scope, it looks in the parent of that scope.

If we create a function A within function B, the creation scope of function A is function B. For example, see the following code snippet:

```
> x <- 10
> wrapper <- function(y){
     x <- 99
     c<- function(y){
          print(x + y)
     }
     return(c)
}
> t <- wrapper()
> t(1)
[1] 100
> x
[1] 10
```

We created a variable x in the global space and gave it a value of 10. We created a function, named it wrapper, and had it accept an argument named y. Within the wrapper() function, we created another variable named x and gave it a value of 99. We also created a function named c. The function wrapper() passes the argument y into the function c(), and the c() function outputs the value of x added to y. Finally, the wrapper() function returns the c() function.

We created a variable t and set it to the returned value of the wrapper() function, which is the function c(). When we run the t() function and pass in a value of 1, we see that it outputs 100 because it is referencing the variable x from the function wrapper().

Being able to reach into the scope of a function that has executed is called a *closure*.

But, you may ask, how can we be sure that we are executing the returned function and not rerunning wrapper() each time? R has a very nice feature where if you type in the name of a function without the parentheses, the interpreter will output the body of the function.

When we do this, we are in fact referencing the returned function and using a closure to reference the x variable:

```
> t
function(y){
        print(x + y)
    }
<environment: 0x17f1d4c4>
```

Summary

In this chapter, we downloaded and installed R. We explored the command line, went over data types, and got up and running importing into the R environment data for analysis. We looked at lists, how to create them, add to them, loop through them, and to apply functions to elements in a list.

We looked at functions, talked about lexical scope, and saw how to create closures in R.

Next chapter, we'll take a deeper dive into R, look at objects, get our feet wet with statistical analysis in R, and explore creating R Markdown documents for distribution over the Web.

CHAPTER 3

A Deeper Dive into R

The last chapter explored some introductory concepts in R, from using the console to importing data. We installed packages and discussed data types, including different list types. We finished up by talking about functions and creating closures.

This chapter will look at object-oriented concepts in R, explore concepts in statistical analysis, and finally see how R can be incorporated into R Markdown for real-time distribution.

Object-Oriented Programming in R

R supports two different systems for creating objects: the S3 and S4 methods. S3 is the default way that objects are handled in R. We've been using and making S3 objects with everything that we've done so far. S4 is a newer way to create objects in R that has more built-in validation, but more overhead. Let's take a look at both methods.

OK, so traditional, class-based, object-oriented design is characterized by creating classes that are the blueprint for instantiated objects (see Figure 3-1).

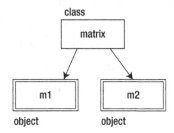

Figure 3-1. *The matrix class is used to create the variables* m1 *and* m2, *both matrices*

© Tom Barker, Jon Westfall 2022
T. Barker and J. Westfall, *Pro Data Visualization Using R and JavaScript*,
https://doi.org/10.1007/978-1-4842-7202-2_3

At a very high level, in traditional object-oriented languages, classes can extend other classes to inherit the parent class' behavior, and classes can also implement interfaces, which are contracts defining what the public signature of the object should be. See Figure 3-2 for an example of this, in which we create an IUser interface that describes what the public interface should be for any user type class, and a BaseUser class that implements the interface and provides a base functionality. In some languages, we might make BaseUser an abstract class, a class that can be extended but not directly instantiated. The User and SuperUser classes extend BaseClass and customize the existing functionality for their own purposes.

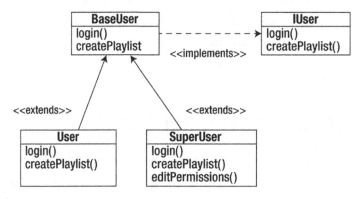

Figure 3-2. *An IUser interface implemented by a superclass BaseUser that the subclasses User and SuperUser extend*

There also exists the concept of polymorphism, in which we can change functionality via the inheritance chain. Specifically, we would inherit a function from a base class but override it, keep the signature (the function name, the type and amount of parameters it accepts, and the type of data that it returns) the same, but change what the function does. Compare overriding a function to the contrasting concept of overloading a function, in which the function would have the same name but a different signature and functionality.

S3 Classes

S3, so called because it was first implemented in version 3 of the S language, uses a concept called **generic functions**. Everything in R is an object, and each object has a string property called class that signifies what the object is. There is no validation around it, and we can overwrite the `class` property ad hoc. That's the main problem with S3—the lack of validation. If you ever had an esoteric error message returned when trying to use a function, you probably experienced the repercussions of this lack of validation firsthand. The error message was probably generated not from R detecting that an incorrect type had been passed in, but from the function trying to execute with what was passed in and failing at some step along the way.

See the following code, in which we create a matrix and change its class to be a vector:

```
> m <- matrix(c(1:10), nrow=2)
> m
     [,1] [,2] [,3] [,4] [,5]
[1,]    1    3    5    7    9
[2,]    2    4    6    8   10
> class(m) <- "vector"
> m
     [,1] [,2] [,3] [,4] [,5]
[1,]    1    3    5    7    9
[2,]    2    4    6    8   10
attr(,"class")
[1] "vector"
```

Generic functions are objects that check the `class` property of objects passed into them and exhibit different behavior based on that attribute. It's a nice way to implement polymorphism. We can see the methods that a generic function uses by passing the generic function to the `methods()` function. The following code shows the methods of the `plot()` generic function:

```
> methods(plot)
 [1] plot.acf*            plot.data.frame*     plot.decomposed.ts*
         plot.default         plot.dendrogram*
 [6] plot.density         plot.ecdf            plot.factor*
         plot.formula*        plot.function
[11] plot.hclust*         plot.histogram*      plot.HoltWinters*
         plot.isoreg*         plot.lm
[16] plot.medpolish*      plot.mlm             plot.ppr*
         plot.prcomp*         plot.princomp*
```

```
[21] plot.profile.nls*    plot.spec            plot.stepfun
        plot.stl*             plot.table*
[26] plot.ts                 plot.tskernel*       plot.TukeyHSD
Non-visible functions are asterisked
```

Notice that within the generic plot() function is a myriad of methods to handle all the different types of data that could be passed to it, such as plot.data.frame for when we pass a data frame to plot(); or if we want to plot a TukeyHSD object plot(), plot.TukeyHSD is ready for us.

Note   Type **?TukeyHSD** for more information on this object.

Now that you know how S3 object-oriented concepts work in R, let's see how to create our own custom S3 objects and generic functions.

An S3 class is a list of properties and functions with an attribute named class. The class attribute tells generic functions how to treat objects that implement a particular class. Let's create an example using the UserClass idea from Figure 3-2:

```
> tom <- list(userid = "tbarker", password = "password123",
playlist=c(12,332,45))
> class(tom) <- "user"
```

We can inspect our new object by using the attributes() function, which tells us the properties that the object has as well as its class:

```
> attributes(tom)
$names
[1] "userid"   "password" "playlist"
$class
[1] "user"
```

Now to create generic functions that we can use with our new class, start by creating a function that will handle only our user object; then generalize it so any class can use it. It will be the createPlaylist() function, and it will accept the user on which to perform the operation and a playlist to set. The syntax for this is [function name].[class name]. Note that we access the properties of S3 objects using the dollar sign.

```
>createPlaylist.user <- function(user, playlist=NULL){
    user$playlist <- playlist
    return(user)
}
```

Note that while you type directly into the console, R enables you to span several lines without executing your input until you complete an expression. After your expression is complete, it will be interpreted. If you want to execute several expressions at once, you can copy and paste into the command line.

Let's test it to make sure it works as desired. It should set the playlist property of the passed-in object to the vector that is passed in:

```
> tom <- createPlaylist.user(tom, c(11,12))
> tom
$userid
[1] "tbarker"
$password
[1] "password123"
$playlist
[1] 11 12
attr(,"class")
[1] "user"
```

Excellent! Now let's generalize the createPlaylist() function to be a generic function. To do this, we just create a function named createPlaylist and have it accept an object and a value. Within our function, we use the UseMethod() function to delegate functionality to our class-specific createPlaylist() function: createPlaylist.user.

The UseMethod() function is the core of generic functions: it evaluates the object, determines its class, and dispatches to the correct class-specific function:

```
> createPlaylist <- function(object, value)
{
    UseMethod("createPlaylist", object)
}
```

Now let's try it out to see whether it worked:

```
> tom <- createPlaylist(tom, c(21,31))
> tom
$userid
[1] "tbarker"
$password
[1] "password123"
$playlist
[1] 21 31
attr(,"class")
[1] "user"
```

Excellent!

S4 Classes

Let's look at S4 objects. Remember that the main complaint about S3 is the lack of validation. S4 addresses this lack by having overhead built into the class structure. Let's take a look.

First, we'll create the user class. We do this with the setClass() function.

- The first parameter in the setClass() function is a string that signifies the name of the class that we are creating.

- The next parameter is called representation, and it is a list of named properties.

```
setClass("user",
representation(userid="character",
    password="character",
    playlist="vector"
)
)
```

We can test it by creating a new object from this class. We use the new() function to create a new instance of the class:

```
> lynn <- new("user", userid="lynn", password="test", playlist=c(1,2))
> lynn
An object of class "user"
Slot "userid":
[1] "lynn"
Slot "password":
[1] "test"
Slot "playlist":
[1] 1 2
```

Very nice. Note that for S4 objects, we use the @ symbol to reference properties of objects:

```
> lynn@playlist
[1] 1 2
> class(lynn)
[1] "user"
attr(,"package")
[1] ".GlobalEnv
```

Let's create a generic function for this class by using the setMethod() function. We simply pass in the function name, the class name, and then an anonymous function that will serve as the generic function:

```
> setMethod("createPlaylist", "user", function(object, value){
    object@playlist <- value
    return(object)
 })
```

Creating a generic function from function 'createPlaylist' in the global environment
[1] "createPlaylist"
>

Let's try it out:

```
> lynn <- createPlaylist(lynn, c(1001, 400))
> lynn
An object of class "user"
Slot "userid":
[1] "lynn"
Slot "password":
[1] "test"
Slot "playlist":
[1] 1001   400
```

Excellent!

Although some prefer the simplicity and flexibility of the S3 way, some prefer the structure of the S4 method; the choice of S3 or S4 objects is purely one of personal preference. My own preference is for the simplicity of S3, and that is what we will be using for the remainder of the book. Google, in its R Style Guide available at `https://google.github.io/styleguide/Rguide.html`, mirrors my own feelings about S3, saying "Use S3 objects and methods unless there is a strong reason to use S4 objects or methods."

Statistical Analysis with Descriptive Metrics in R

Now let's take a look at some concepts in statistical analysis and how to implement them in R. You might remember most of the concepts covered in this chapter from an introductory statistics class from college; they are the base concepts needed to begin to think about and discuss your data.

First, let's get some data on which we'll perform statistical analysis. R comes preloaded with a number of data sets that we can use as sample data. To see a list of available data sets with your install, simply type `data()` at the console. You'll be presented with the screen that you see in Figure 3-3.

```
Save   Print                                               Functions                                          Search          Q▾ Help search

 1   Data sets in package 'datasets':
 2
 3   AirPassengers                  Monthly Airline Passenger Numbers 1949-1960
 4   BJsales                        Sales Data with Leading Indicator
 5   BJsales.lead (BJsales)         Sales Data with Leading Indicator
 6   BOD                            Biochemical Oxygen Demand
 7   CO2                            Carbon Dioxide Uptake in Grass Plants
 8   ChickWeight                    Weight versus age of chicks on different diets
 9   DNase                          Elisa assay of DNase
10   EuStockMarkets                 Daily Closing Prices of Major European Stock Indices, 1991-1998
11   Formaldehyde                   Determination of Formaldehyde
12   HairEyeColor                   Hair and Eye Color of Statistics Students
13   Harman23.cor                   Harman Example 2.3
14   Harman74.cor                   Harman Example 7.4
15   Indometh                       Pharmacokinetics of Indomethacin
16   InsectSprays                   Effectiveness of Insect Sprays
17   JohnsonJohnson                 Quarterly Earnings per Johnson & Johnson Share
18   LakeHuron                      Level of Lake Huron 1875-1972
19   LifeCycleSavings               Intercountry Life-Cycle Savings Data
20   Loblolly                       Growth of Loblolly pine trees
21   Nile                           Flow of the River Nile
22   Orange                         Growth of Orange Trees
23   OrchardSprays                  Potency of Orchard Sprays
24   PlantGrowth                    Results from an Experiment on Plant Growth
25   Puromycin                      Reaction Velocity of an Enzymatic Reaction
26   Seatbelts                      Road Casualties in Great Britain 1969-84
27   Theoph                         Pharmacokinetics of Theophylline
28   Titanic                        Survival of passengers on the Titanic
29   ToothGrowth                    The Effect of Vitamin C on Tooth Growth in Guinea Pigs
30   UCBAdmissions                  Student Admissions at UC Berkeley
31   UKDriverDeaths                 Road Casualties in Great Britain 1969-84
32   UKgas                          UK Quarterly Gas Consumption
33   USAccDeaths                    Accidental Deaths in the US 1973-1978
34   USArrests                      Violent Crime Rates by US State
35   USJudgeRatings                 Lawyers' Ratings of State Judges in the US Superior Court
36   USPersonalExpenditure          Personal Expenditure Data
37   VADeaths                       Death Rates in Virginia (1940)
38   WWWusage                       Internet Usage per Minute
39   WorldPhones                    The World's Telephones
40   ability.cov                    Ability and Intelligence Tests
41   airmiles                       Passenger Miles on Commercial US Airlines, 1937-1960
42   airquality                     New York Air Quality Measurements
43   anscombe                       Anscombe's Quartet of 'Identical' Simple Linear Regressions
44   attenu                         The Joyner-Boore Attenuation Data
45   attitude                       The Chatterjee-Price Attitude Data
```

Figure 3-3. *Available data sets in R*

To see the contents of a data set, you can call it by name in the console. Let's take a look at the USArrests data set, which we'll use for the next few topics.

```
> USArrests
            Murder  Assault  UrbanPop Rape
Alabama      13.2     236      58      21.2
Alaska       10.0     263      48      44.5
Arizona       8.1     294      80      31.0
Arkansas      8.8     190      50      19.5
California    9.0     276      91      40.6
Colorado      7.9     204      78      38.7
```

Connecticut	3.3	110	77	11.1
Delaware	5.9	238	72	15.8
Florida	15.4	335	80	31.9
Georgia	17.4	211	60	25.8
Hawaii	5.3	46	83	20.2
Idaho	2.6	120	54	14.2
Illinois	10.4	249	83	24.0
Indiana	7.2	113	65	21.0
Iowa	2.2	56	57	11.3
Kansas	6.0	115	66	18.0
Kentucky	9.7	109	52	16.3
Louisiana	15.4	249	66	22.2
Maine	2.1	83	51	7.8
Maryland	11.3	300	67	27.8
Massachusetts	4.4	149	85	16.3
Michigan	12.1	255	74	35.1
Minnesota	2.7	72	66	14.9
Mississippi	16.1	259	44	17.1
Missouri	9.0	178	70	28.2
Montana	6.0	109	53	16.4
Nebraska	4.3	102	62	16.5
Nevada	12.2	252	81	46.0
New Hampshire	2.1	57	56	9.5
New Jersey	7.4	159	89	18.8
New Mexico	11.4	285	70	32.1
New York	11.1	254	86	26.1
North Carolina	13.0	337	45	16.1
North Dakota	0.8	45	44	7.3
Ohio	7.3	120	75	21.4
Oklahoma	6.6	151	68	20.0
Oregon	4.9	159	67	29.3
Pennsylvania	6.3	106	72	14.9
Rhode Island	3.4	174	87	8.3
South Carolina	14.4	279	48	22.5
South Dakota	3.8	86	45	12.8

Tennessee	13.2	188	59	26.9
Texas	12.7	201	80	25.5
Utah	3.2	120	80	22.9
Vermont	2.2	48	32	11.2
Virginia	8.5	156	63	20.7
Washington	4.0	145	73	26.2
West Virginia	5.7	81	39	9.3
Wisconsin	2.6	53	66	10.8
Wyoming	6.8	161	60	15.6

>

The first function in R that we'll look at is the summary() function, which accepts an object and returns the following key descriptive metrics, grouped by column:

- Minimum value

- Maximum value

- Median for numbers and frequency for strings

- Mean

- First quartile

- Third quartile

Let's run the USArrests data set through the summary() function:

```
> summary(USArrests)
     Murder          Assault         UrbanPop          Rape
 Min.   : 0.800   Min.   : 45.0   Min.   :32.00   Min.   : 7.30
 1st Qu.: 4.075   1st Qu.:109.0   1st Qu.:54.50   1st Qu.:15.07
 Median : 7.250   Median :159.0   Median :66.00   Median :20.10
 Mean   : 7.788   Mean   :170.8   Mean   :65.54   Mean   :21.23
 3rd Qu.:11.250   3rd Qu.:249.0   3rd Qu.:77.75   3rd Qu.:26.18
 Max.   :17.400   Max.   :337.0   Max.   :91.00   Max.   :46.00
```

Let's look at each of these metrics in detail, as well as the standard deviation.

Median and Mean

Note that the median is the number that is the middle value in a data set, quite literally the number that has the same amount of numbers greater and less than itself in the data set. If our data set looks like the following, the median is 3:

```
1, 2, 3, 4, 5
```

But notice that it's easy to find the median when there are an odd number of items in a data set. Suppose that there is an even number of items in a data set, as follows:

```
1, 2, 3, 4, 5, 6
```

In this case, we take the middle pair, 3 and 4, and get the average of those two numbers. The median is 3.5.

Why does the median matter? When you look at a data set, there are usually outliers at either end of the spectrum, values that are much higher or much lower than the rest of the data set. Gathering the median value excludes these outliers, giving a much more realistic view of the average values.

Contrast this idea with the mean, which is simply the sum of the values in a data set divided by the number of items. The values include the outliers, so the mean can be skewed by having significant outliers and really represent the full data set.

For example, look at the following data set:

```
1, 2, 3, 4, 30
```

The median is still 3 for this data set, but the mean is 8, because of this:

```
median = [1,2] 3 [4,30]
mean =   1 + 2 + 3 + 4 + 30 = 40
    40 / 5 = 8
```

Quartiles

The median is the center of the data set, which means that the median is the second quartile. Quartiles are the points that divide a data set into four even sections. We can use the quantile() function to pull just the quartiles from our data set.

```
> quantile(USArrests$Murder)
   0%    25%    50%    75%   100%
0.800  4.075  7.250 11.250 17.400
```

The summary() function simply returns the quartiles, as well as the minimum, maximum, and mean values. Here are the summary() results for comparison, with the previous quartiles highlighted:

```
> summary(USArrests)
     Murder          Assault         UrbanPop          Rape
 Min.   : 0.800   Min.   : 45.0   Min.   :32.00   Min.   : 7.30
 1st Qu.: 4.075   1st Qu.:109.0   1st Qu.:54.50   1st Qu.:15.07
 Median : 7.250   Median :159.0   Median :66.00   Median :20.10
 Mean   : 7.788   Mean   :170.8   Mean   :65.54   Mean   :21.23
 3rd Qu.:11.250   3rd Qu.:249.0   3rd Qu.:77.75   3rd Qu.:26.18
 Max.   :17.400   Max.   :337.0   Max.   :91.00   Max.   :46.00
```

Standard Deviation

Speaking of the idea of the mean, there is also the idea that data has a normal distribution or that data is normally densely clustered around the mean with lighter groupings above and below the mean. This is usually demonstrated with a bell curve, in which the mean is the top of the curve and the outliers are distributed on either end of it (see Figure 3-4).

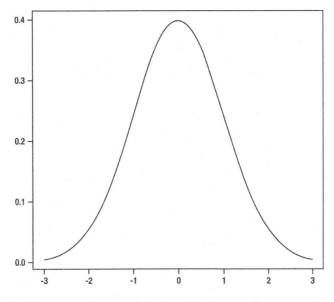

Figure 3-4. *The bell curve of a normal distribution*

Standard deviation is a unit of measurement that describes the average of how far apart the data is distributed from the mean, so we can detail how far each data point is from the mean in terms of standard deviations.

In R, we can determine the standard deviation using the sd() function. The sd() function expects a vector of numeric values:

```
> sd(USArrests$Murder)
[1] 4.35551
```

If we want to gather the standard deviation for a matrix, we can use the sapply() function to apply the sd() function, like so:

```
> sapply(USArrests, sd)
  Murder     Assault    UrbanPop         Rape murderRank
4.355510   83.337661   14.474763     9.366385   14.574930
```

RStudio IDE

If you prefer to develop in an integrated development environment (IDE) instead of at the command line, you can use a free product called RStudio IDE. The RStudio IDE is made by the RStudio company and is much more than just an IDE (as you will soon see). The RStudio company was founded by JJ Allaire, creator of ColdFusion. RStudio IDE is available for download at www.rstudio.com/ide/ (see Figure 3-5 for a screenshot of the download page).

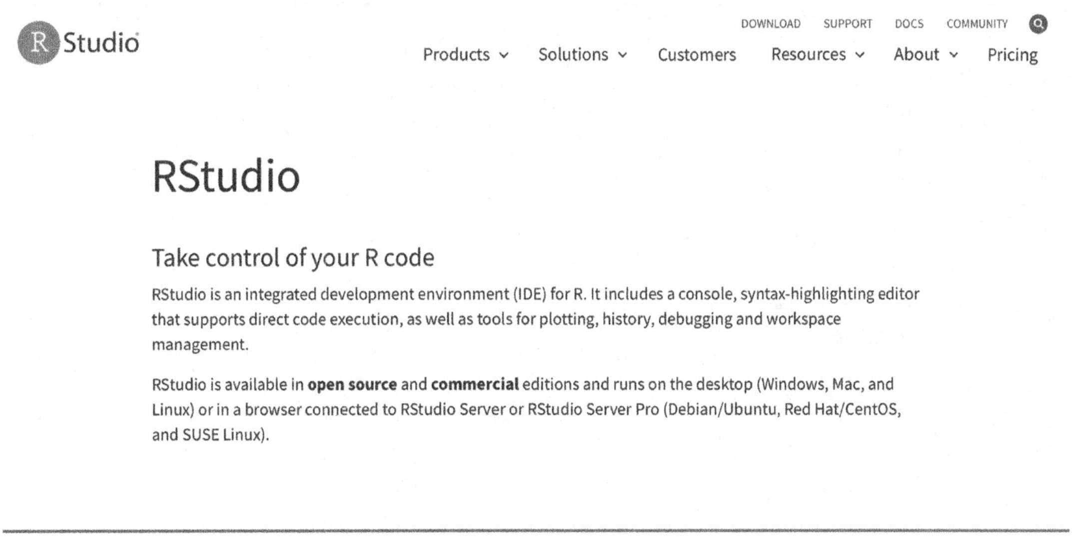

 Studio

DOWNLOAD   SUPPORT   DOCS   COMMUNITY

Products ∨   Solutions ∨   Customers   Resources ∨   About ∨   Pricing

RStudio

Take control of your R code

RStudio is an integrated development environment (IDE) for R. It includes a console, syntax-highlighting editor that supports direct code execution, as well as tools for plotting, history, debugging and workspace management.

RStudio is available in **open source** and **commercial** editions and runs on the desktop (Windows, Mac, and Linux) or in a browser connected to RStudio Server or RStudio Server Pro (Debian/Ubuntu, Red Hat/CentOS, and SUSE Linux).

Figure 3-5. *RStudio IDE home page*

Note   You should install the RStudio IDE now because you will use it in the remainder of this chapter.

After installation, the IDE is split into four panes (see Figure 3-6).

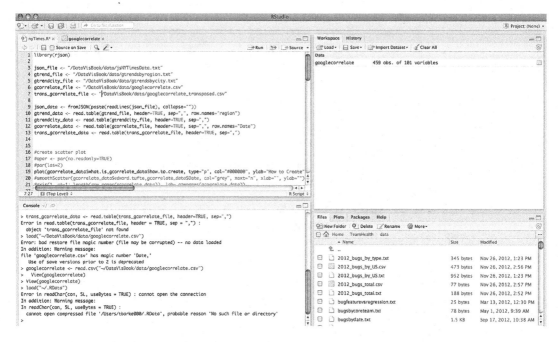

Figure 3-6. *RStudio Interface*

The upper-left pane is the R script file in which we edit our R source code. The bottom-left pane is the R command line. The upper-right side pane holds the command history as well as all the objects in our current workspace. The bottom-right pane is split into tabs that can show the following:

- Contents of the file system for the current working directory

- Plots or charts that have been generated

- Current packages installed

- R help pages

Although it is great to have everything that you need in one place, here is where things become really interesting.

R Markdown

In version 0.96 of RStudio, the team announced support for R Markdown using the knitr package. We can now embed R code into markdown documents that can get interpreted by knitr into HTML (HyperText Markup Language). But it gets even better.

The RStudio company also makes a product called RPubs that allows users to create accounts and host their R Markdown files for distribution over the Web.

Note   Markdown is a plain text markup language created by John Gruber and Aaron Swartz. In markdown, you can use simple and lightweight text encodings to signify formatting. The markdown document is read and interpreted and an HTML file is output.

A quick overview of markdown syntax follows:

```
header 1
=========
header 2
--------------
###header 3
####header 4
*italic*
**bold**
[link text]([URL])
![alt text]([path to image])
```

The great thing about R Markdown is that we can embed R code within our markdown document. We embed R using three tick marks and the letter r in curly braces:

```
```{r}
[R code]
```
```

We need three things to begin creating R Markdown (.rmd) documents:

- R
- R Studio IDE version 0.96 or higher
- The knitr package

The knitr package is used to reformat R into several different output formats, including HTML, markdown, or even plain text.

Note   Information about the knitr package is available at `http://yihui.name/knitr/`.

Because you already have R and RStudio IDE installed, you will first install knitr. R Studio IDE has a nice interface to install packages: simply go to the Tools file menu, and click Install Packages. You should see the pop-up that is shown in Figure 3-7, in which you can specify the package name (R Studio IDE has a nice type ahead here for package discovery) and what library to install to.

Figure 3-7. *Installing the knitr package*

After knitr is installed, you need to close and relaunch RStudio IDE. You then go to the File menu, and choose File ➤ New, in which you should see a number of options, including R Markdown. If you choose R Markdown, and choose the default option of "Document" and "HTML" as the Default Output Format, you get a new file with the template shown in Figure 3-8.

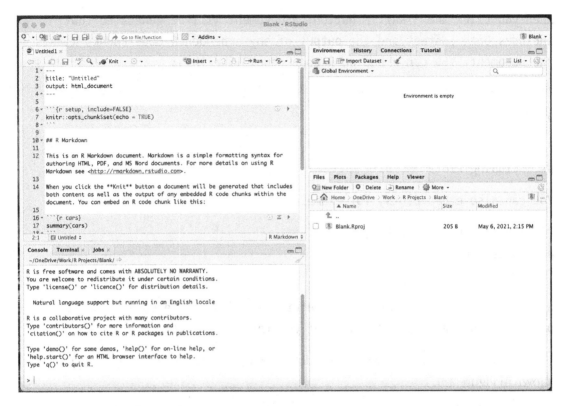

Figure 3-8. *RStudio IDE*

The R Markdown template has the following code:

```
---
title: "Untitled"
output: html_document
---

```{r setup, include=FALSE}
knitr::opts_chunk$set(echo = TRUE)
```

R Markdown

This is an R Markdown document. Markdown is a simple formatting syntax for
authoring HTML, PDF, and MS Word documents. For more details on using R
Markdown, see <http://rmarkdown.rstudio.com>.
```

When you click the **Knit** button, a document will be generated that includes both content and the output of any embedded R code chunks within the document. You can embed an R code chunk like this:

```
```{r cars}
summary(cars)
```
```

Including Plots

You can also embed plots, for example:

```
```{r pressure, echo=FALSE}
plot(pressure)
```
```

Note that the `echo = FALSE` parameter was added to the code chunk to prevent printing of the R code that generated the plot.

This is the template, and when you click the Knit button, you will see the output shown in Figure 3-9.

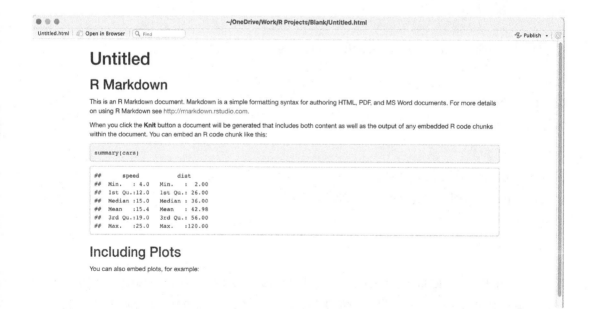

Figure 3-9. *HTML output of RMarkdown template*

Did you notice the Publish button at the top of Figure 3-9? That is how we push our R Markdown file to RPubs for hosting and distribution over the Web.

RPubs

RPubs is a free web publishing platform for R Markdown files, provided by RStudio (the company). You can create a free account by visiting `www.rpubs.com`. Figure 3-10 shows a screenshot of the RPubs home page.

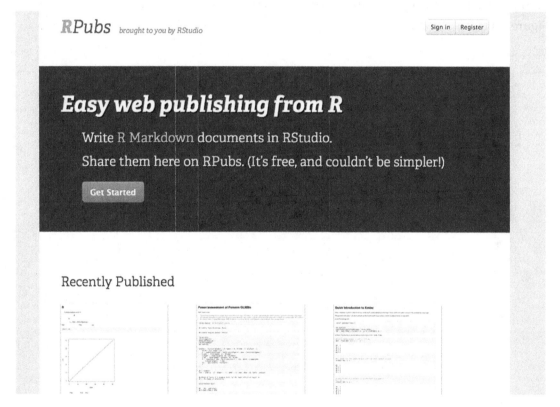

Figure 3-10. *RPubs home page*

Just click the Register button, and fill out the form to create your free account. RPubs is fantastic; it's a platform in which we can post our R Markdown documents for distribution.

Caution   Be aware that every file you put up on RPubs is publicly available, so be sure not to put any sensitive or proprietary information in it. If you don't want to put your R Markdown files where they are available for the whole world to see, you can instead click the Save As button right next to the Publish button to save the file as regular HTML.

After you click the Publish button, you are prompted to log in with your RPubs account. After logging in, you will be directed to the Document Details page, as seen in Figure 3-11.

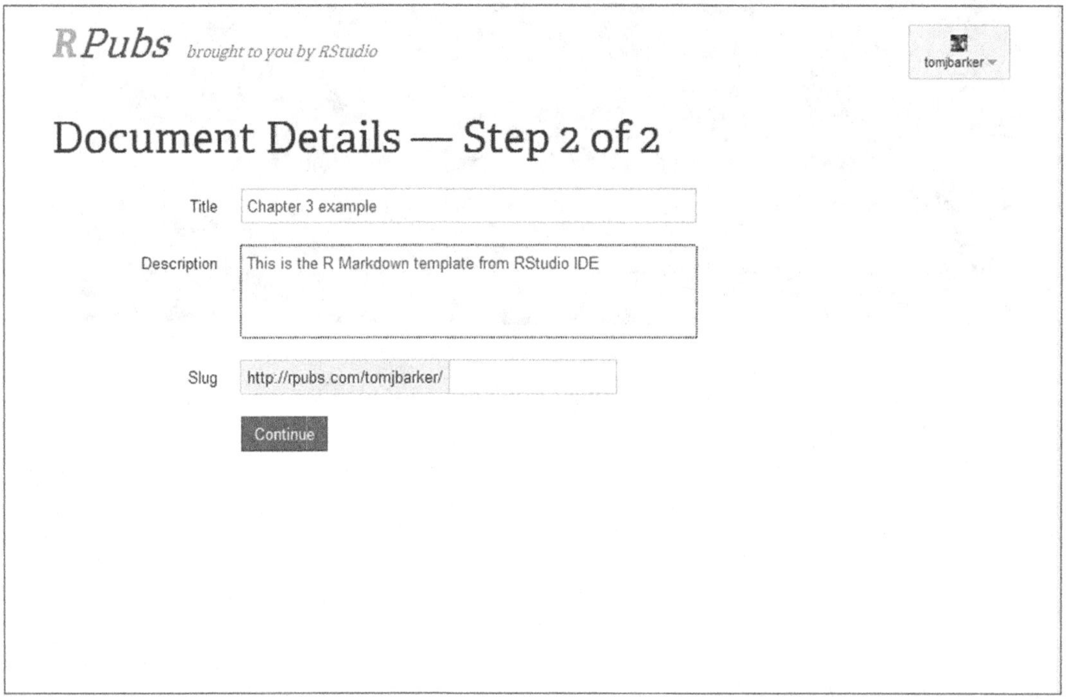

Figure 3-11. *Publishing to RPubs*

After filling out the document details, a title for your document, and a description, you will be directed to your document hosted in RPubs. See Figure 3-12 for the template from Figure 3-9 hosted in RPubs and available publicly here: www.rpubs.com/tomjbarker/3370.

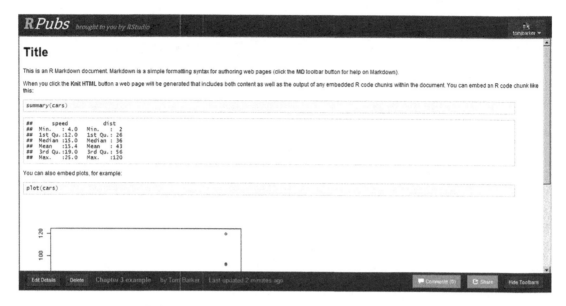

Figure 3-12. *RMarkdown template published to RPubs*

This is a powerful distribution method for R documents and for communicating data visualizations. In the coming chapters, we will put all the completed R charts up on RPubs for public consumption.

Summary

This chapter explored some deeper concepts in R, from the different models of object-oriented design available to how to do statistical analysis with R. We even looked at how to use RMarkdown and RPubs to make data visualizations in R available for public distribution.

In the next chapter, we will look at D3, a JavaScript library that enables us to analyze and visualize data within the browser and add interactivity to visualizations.

CHAPTER 4

Data Visualization with D3

Thus far, when we have been talking about technologies used to create data visualizations, we've been talking about R. We've spent the last two chapters exploring the R environment and learning about the command line. We covered introductory topics in the R language, ranging from data types, functions, and object-oriented programming. We even talked about how to publish our R documents to the Web using RPubs.

This chapter we will look at a JavaScript library called D3 that is used to create interactive data visualizations. First is a very quick primer on HTML, CSS, and JavaScript, the supporting languages of D3, to level set. Then we'll dig into D3 and explore how to make some of the more commonly used charts in D3.

Preliminary Concepts

D3 is a JavaScript library. Specifically, that means it is written in JavaScript and embedded in an HTML page. We can reference the objects and functions in D3 in our own JavaScript code. So let's start at the beginning. The purpose of the next section is not to take a deep dive into HTML CSS and JavaScript; there are plenty of other resources for that, including *Foundation Website Creation* that I helped to co-write. The purpose is to have a very high-level recap of concepts that we will deal with directly with D3. If you are already familiar with HTML, CSS, and JavaScript, you can skip down to the "History of D3" section of this chapter.

© Tom Barker, Jon Westfall 2022
T. Barker and J. Westfall, *Pro Data Visualization Using R and JavaScript*,
https://doi.org/10.1007/978-1-4842-7202-2_4

HTML

HTML is a markup language; in fact, it stands for HyperText Markup Language. It is a presentation language, made up of elements that signify formatting and layout. Elements contain attributes that have values that specify details about the element, tags, and content. To explain, let's look at our basic HTML skeletal structure that we will use for most of our examples in this chapter:

```
<!DOCTYPE html>
<html>
<head></head>
<body></body>
</html>
```

Let's start at the first line. That is the doctype that tells the browser's render engine what rule set to use. Browsers can support multiple versions of HTML, and each version has a slightly different rule set. The doctype specified here is the HTML5 doctype. Another example of a doctype is this:

```
<!DOCTYPE html PUBLIC "-//W3C//DTD XHTML 1.1//EN""  http://www.w3.org/TR/
xhtml11/DTD/xhtml11.dtd  ">
```

This is the doctype for XHTML 1.1. Notice that it specifies the URL of the document type definition (.dtd). If we were to read the plain text at the URL, we would see that it is a specification for how to parse HTML tags. The W3C maintains a list of doctypes here: www.w3.org/QA/2002/04/valid-dtd-list.html.

MODERN BROWSER ARCHITECTURE

Modern browsers are composed of modular pieces that encapsulate very specific functionality. These modules can also be licensed out and embedded in other applications:

- They have a UI layer that handles drawing the user interface of the browser, like the window, the status bar, and the back button.

- They have render engines to parse, tokenize, and paint the HTML.

- They have a network layer to handle the network operations involved in retrieving the HTML and all the assets on the page.

- They have a JavaScript engine to interpret and execute the JavaScript in the page.

See Figure 4-1 for a representation of this architecture.

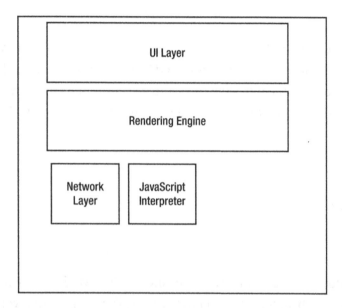

Figure 4-1. *Modern Browser Architecture*

Back to the skeletal HTML structure. The next line is the `<html>` tag; this is the root-level tag for the document and holds every other HTML element that we will use. Notice that there is a closing tag on the last line of the document.

Next is the `<head>` tag, which is a container that generally holds information that is not displayed on the page (e.g., the title and meta-information). After the `<head>` tag is the `<body>` tag, which is a container that holds all the HTML elements that will be displayed on the page, for example, paragraphs:

```
<p> this is a paragraph </p>
```

or links:

```
<a href="[URL]">link text or image here</a>
```

or images:

```
<img src="[URL]"/>
```

When it comes to D3, most of the JavaScript that we will be writing will be in the body section, and most of the CSS will be in the head section.

CSS

CSS stands for Cascading Style Sheets and is what is used to style the HTML elements on a web page. Style sheets are either contained in `<style>` tags or linked externally via `<link>` tags and are comprised of style rules and selectors. Selectors target the element on the web page to style, and the style rule defines what styles to apply. Let's look at an example:

```
<style>
p{
    color: #AAAAAA;
}
</style>
```

In the preceding code snippet, the style sheet is in a `style` tag. The `p` is the **selector** that tells the browser to target every paragraph tag on the web page. The style rule is wrapped in curly braces and is made up of properties and values. This case sets the color of the text in all the paragraphs to #AAAAAA which is the hexadecimal value of a light gray.

Selectors are where the real nuance of CSS is. This is relevant to us because D3 also uses CSS selectors to target elements. Similar to how S3/S4 classes can inherit from each other in R, we can get very specific with selectors and target elements by class or id, or we can use pseudo-classes to target abstract concepts such as when an element is hovered over. We can target ancestors and descendants of elements, up and down the DOM.

Note   The DOM stands for the Document Object Model and is the application programming interface (API) that allows JavaScript to interact with the HTML elements that are on a web page.

```
.classname{
/* style sheet for a class*/
}
#id{
/*style sheet for an id*/
}
element:pseudo-class{
}
```

SVG

The next introductory concept for D3 is SVG, which stands for Scalable Vector Graphics. SVG, which is a standardized way to create vector graphics in the browser, is what D3 uses to create data visualizations. The core functionality that we are concerned about in SVG is the capability to draw shapes and text and integrate them into the DOM so that our shapes can be scripted via JavaScript.

Note   Vector graphics are graphics that are created using points and lines that are mathematically calculated and displayed by the rendering engine. Contrast this idea with bitmap or raster graphics in which the pixel display is prerendered. Vectors, as they are simply equations, tend to scale better and are smaller. However, they lack the depth that bitmap or raster graphics will have.

SVG is essentially its own markup language with its own doctype. We can write SVG in external .svg files or include the SVG tags directly in our HTML. Writing the SVG tags in our HTML page allows us to interact with our shapes via JavaScript.

SVG has support for predefined shapes as well as the capability to draw lines. The predefined shapes in SVG are these:

- `<rect>` to draw rectangles

- `<circle>` to draw circles

- `<ellipse>` to draw ellipses

- `<line>` to draw lines; also `<polyline>` and `<polygon>` to draw lines with multiple points

Let's look at some code examples. If we will write our SVG into an HTML document, we use the `<svg>` tag to wrap our shapes. The `<svg>` takes the `xmlns` and `version` attributes. The `xmlns` attribute should be the path to the SVG namespace, and the `version` is obviously the version of SVG:

```
<svg xmlns=" http://www.w3.org/2000/svg " version="1.1">
</svg>
```

If we are writing stand-alone `.svg` files, we include the full `doctype` and `xml` tags to the page file:

```
<?xml version="1.0" standalone="no"?>
<!DOCTYPE svg PUBLIC "-//W3C//DTD SVG 1.1//EN" " http://www.w3.org/
Graphics/SVG/1.1/DTD/svg11.dtd ">
<svg xmlns=" http://www.w3.org/2000/svg " version="1.1">
</svg>
```

Either way, we create our shapes within the `<svg>` tag. Let's create some sample shapes in our `<svg>` tag:

```
<svg xmlns=" http://www.w3.org/2000/svg " version="1.1" viewBox="0 0 500 500">
    <rect x="10" y="10" width="10" height="100" stroke="#000000"
    fill="#AAAAAA" />
    <circle cx="70" cy="50" r="40" stroke="#000000" fill="#AAAAAA" />
    <ellipse cx="230" cy="60" rx="100" ry="50" stroke="#000000"
    fill="#AAAAAA" />
</svg>
```

This code produces the shapes shown in Figure 4-2.

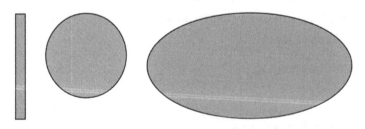

Figure 4-2. *A rectangle, circle, and ellipse drawn in SVG*

Notice that we assign x and y coordinates for all the shapes—in the case of the circle and ellipse cx and cy coordinates—as well as fill color and stroke colors. This is just the smallest taste; we can also create gradients and filters and then apply them to our shapes. We can also create text to use in our SVG drawings using the <text> tag.

Let's take a look. We'll update the preceding SVG code to add text labels for each shape:

```
<svg xmlns=" http://www.w3.org/2000/svg " version="1.1" viewBox="0 0
500 500">
    <rect x="80" y="20" width="10" height="100" stroke="#000000"
    fill="#AAAAAA" />
    <text x="55" y="145" fill="#000000">rectangle</text>
    <circle cx="170" cy="60" r="40" stroke="#000000" fill="#AAAAAA" />
    <text x="150" y="145" fill="#000000">circle</text>
    <ellipse cx="330" cy="70" rx="100" ry="50" stroke="#000000"
    fill="#AAAAAA" />
    <text x="295" y="145" fill="#000000">ellipse</text>
</svg>
```

This code creates the drawing shown in Figure 4-3.

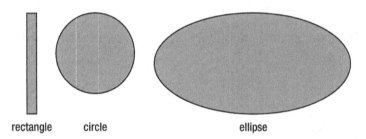

Figure 4-3. *SVG shapes with text labels*

Now we can start to see the possibilities of creating data visualizations with just these fundamental building blocks. Because D3 is a JavaScript library, and most of the work we will be doing with D3 will be in JavaScript, let's next take a high-level look at JavaScript before we delve into D3.

JavaScript

JavaScript is the scripting language of the Web. JavaScript can be included in an HTML document either by placing `script` tags inline in the document or by linking to an external JavaScript document:

```
<script>
//javascript goes here
</script>
<script src="pathto.js"></script>
```

JavaScript can be used to process information, react to events, and interact with the DOM. In JavaScript, we create variables using the `var` keyword.

```
var foo = "bar";
```

Note that if we do not use the `var` keyword, the variable that we create is assigned to the global scope. We don't want to do this because our globally scoped variable could then be overwritten by any other code on our web page.

JavaScript looks much like other C-based languages in that each expression ends in a semicolon, and blocks of code such as function and conditional bodies are wrapped in curly braces.

Conditional statements are generally `if-else` statements formatted as follows:

```
if([condition]){
    [code to execute]
}else{
    [code to execute]
}
```

Functions are formatted like so:

```
function [function name] ([arguments]){
    [code to execute]
}
```

We access DOM elements in JavaScript usually by referencing the element by its `id` attribute. We do this like using the `getElementById()` function:

```
var header = document.getElementById("header");
```

The preceding code stores a reference to the element on the web page that has an ID of header. We can then update properties of this element, including adding new elements or removing the element altogether.

Objects in JavaScript are generally object literals, meaning that we craft them at runtime, composed of properties and methods. We create object literals like so:

```
var myObj = {
    myProp: 20,
    myfunc: function(){
    }
}
```

We reference properties and methods of objects using the dot operator:

```
myObj.myprop = 10;
```

See, that was fast and painless. OK, on to D3!

History of D3

D3 stands for Data-Driven Documents and is a JavaScript library used to create interactive data visualizations. The seed of the idea that would become D3 started in 2009 as Protovis, created by Mike Bostock, Vadim Ogievetsky, and Jeff Heer while they were with the Stanford Visualization Group.

Note Information on the Stanford Visualization Group can be found at its website: http://vis.stanford.edu/. The original white paper for Protovis can be found at http://vis.stanford.edu/papers/protovis.

Protovis was a JavaScript library that provided an interface for creating different types of visualizations. The root namespace was pv, and it provided an API for creating bars and dots and areas, among other things. Like D3, Protovis used SVG to create these shapes, but unlike D3, it wrapped the SVG calls in its own proprietary nomenclature.

Protovis was abandoned in 2011, so its creators could take their learning and instead create and focus on D3. There is a difference in philosophy between Protovis and D3. Where Protovis aimed to provide wrapped functionality for creating data visualizations, D3 instead facilitates and streamlines the creation of data visualization by working with existing web standards and nomenclature. In D3, we create rectangles and circles in SVG, just facilitated by the syntactic sugar of D3.

Using D3

The first thing we need to do to get working with D3 is to go to the D3 website, `http://d3js.org/`, and download the latest version (see Figure 4-4).

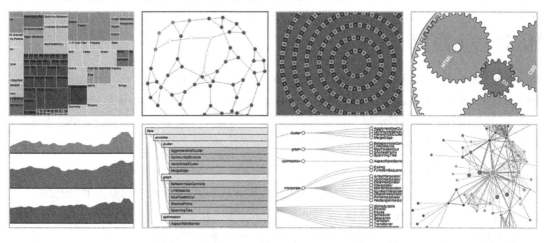

Figure 4-4. *D3 home page*

After that is installed, you can set up a project.

Setting Up a Project

We can include the `.js` file directly on our page, like so:

```
<script src="d3.v3.js"></script>
```

The root namespace is d3; all the commands that we issue from D3 will be using the d3 object.

Using D3

We use the `select()` function to target specific elements or the `selectAll()` function to target all of a specific element type:

```
var body = d3.select("body");
```

The previous line selects the body tag and stores it in a variable named body. We can then change attributes of the body if we want to or add new elements to the body:

```
var allParagraphs = d3.select("body").selectAll("p");
```

The previous line selects the body tag and then selects all the paragraph tags within the body.

Notice that we chained the two actions together on the second line? We selected the body and then selected all the paragraphs, both actions chained together. Also note that we used the CSS selector to specify the element to target.

OK, once we have selected an element, that is now considered our selection and we can perform actions on that selection. We can select elements within our selection as we did in the previous example.

We can update attributes of the selection with the `attr()` function. The `attr()` function accepts two parameters: the first is the name of the attribute, and the second is the value to set the attribute to. Suppose we want to change the background color of the current document. We can just select the body and set the `bgcolor` attribute by adding this to our script block:

```
<script>
    d3.select("body")
        .attr("bgcolor", "#000000");
</script>
```

Notice in the previous code snippet that we have brought the chained attribute function call to the next line. We have done this for readability.

The really fun thing with this is that because we're talking about JavaScript, and functions are first-class objects in JavaScript, we can pass in a function as the value of an attribute so that whatever it evaluates to becomes the value that is set:

```
<script>
    d3.select("body")
        .attr("bgcolor", function(){
        return "#000000";
});
</script>
```

We can also add elements to our selection using the append() function. The append() function accepts a tag name as the first parameter. It will create a new element of the type specified and return that new element as the current selection:

```
<script>
var svg = d3.select("body")
        .append("svg");
</script>
```

The preceding code creates a new SVG tag in the body of the page and stores that selection in the variable svg.

Next, let's re-create the shapes in Figure 4-3 using what we've just learned about D3:

```
<script>
    var svg = d3.select("body")
        .append("svg")
        .attr("width", 800);
    var r = svg.append("rect")
        .attr("x", 80)
        .attr("y", 20)
        .attr("height", 100)
        .attr("width", 10)
        .attr("stroke", "#000000")
        .attr("fill", "#AAAAAA");
```

```
    var c = svg.append("circle")
        .attr("cx", 170)
        .attr("cy", 60)
        .attr("r", 40)
        .attr("stroke", "#000000")
        .attr("fill", "#AAAAAA");
    var e = svg.append("ellipse")
        .attr("cx", 330)
        .attr("cy", 70)
        .attr("rx", 100)
        .attr("ry", 50)
        .attr("stroke", "#000000")
        .attr("fill", "#AAAAAA");
</script>
```

For each shape, we append a new element to the SVG element and update the attributes.

If we compare the two methods, we can see that we just create the SVG element in D3, just as we do in straight markup. We then create an SVG rectangle, circle, and ellipse inside the SVG element along with the same attributes that we specified in the SVG markup. But our D3 example has one very important difference: we now have references to each element on the page that we can interact with.

Let's take a look at interactions in D3.

Binding Data

For data visualizations, the most important interaction we have with our SVG shapes is to bind data to them. This allows us to then reflect that data in the properties of the shapes.

To bind data, we simply call the data() method of a selection:

```
<script>
var rect = svg
    .append("rect")
    .data([1,2,3]);
</script>
```

That's fairly straightforward. We can then reference that bound data via anonymous functions that we pass to our `attr()` function calls. Let's take a look at an example.

First, let's create an array that we will call `dataSet`. To start to envision how this will correlate to creating a data visualization, you can think of `dataSet` as a list of nonsequential values, maybe test scores for a class or total rainfall for a set of regions:

```
<script>
var dataSet = [ 84,62,40,109];
</script>
```

Next, we will create an SVG element on the page. To do that, we'll select the body and append an SVG element with a width of 800 pixels. We'll keep a reference to this SVG element in a variable called `svg`:

```
<script>
var svg = d3
    .select("body")
    .append("svg")
    .attr("width", 800);
</script>
```

Here is where being able to bind data changes things. We will chain together a series of commands that will create placeholder rectangles in the SVG element based on how many elements exist in our data array.

We will first use `selectAll()` to return a reference to all rectangles in the SVG element. There are none yet, but there will be by the time the chain finishes executing. Next in the chain, we bind our `dataSet` variable and call `enter()`. The `enter()` function creates placeholder objects from the bound data. Finally, we call `append()` to create a rectangle at each placeholder that `enter()` created.

```
<script>
bars = svg
    .selectAll("rect")
    .data(dataSet)
    .enter()
    .append("rect");
</script>
```

If we looked at our work so far in a browser, we would see a blank page, but if we looked at the HTML in a web inspector such as Firebug, we would see the SVG element along with the rectangles created, but with no styling or attributes specified yet, similar to Figure 4-5.

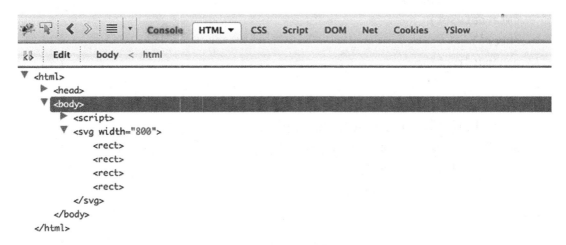

Figure 4-5. *Firebug Inspection interface*

Next, let's style the rectangles that we just made. We have a reference to all the rectangles in the variable bars, so let's chain together a bunch of attr() calls to style the rectangles. While we're at it, let's use our bound data to size the height of the bars.

```
<script>
bars
    .attr("width", 15 )
    .attr("height", function(x){return x;})
    .attr("x", function(x){return x + 40;})
    .attr("fill", "#AAAAAA")
    .attr("stroke", "#000000");
</script>
```

The full source code looks like the following and makes the shapes that we see in Figure 4-6:

```
<script>
var dataSet = [84,62,40,109];
var svg = d3
```

```
    .select("body")
    .append("svg")
    .attr("width", 800);
bars = svg
    .selectAll("rect")
    .data(dataSet)
    .enter()
    .append("rect");
bars
    .attr("width", 15 )
    .attr("height", function(x){return x;})
    .attr("x", function(x){return x + 40;})
    .attr("fill", "#AAAAAA")
    .attr("stroke", "#000000");
</script>
```

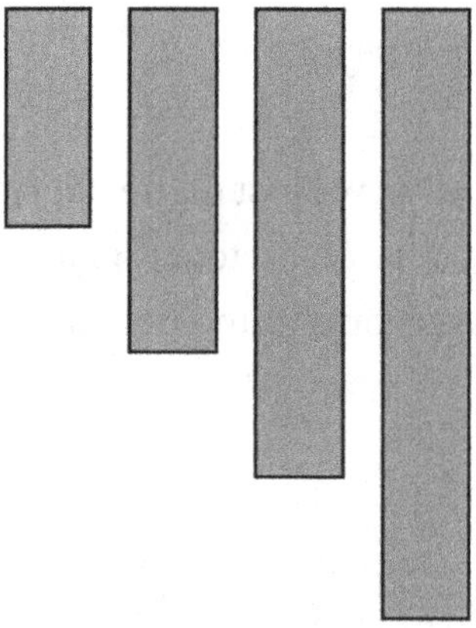

Figure 4-6. *Styled rectangles for bar chart*

Now look in Firebug or your browser's debugging tools again; you can see the generated markup, as shown in Figure 4-7.

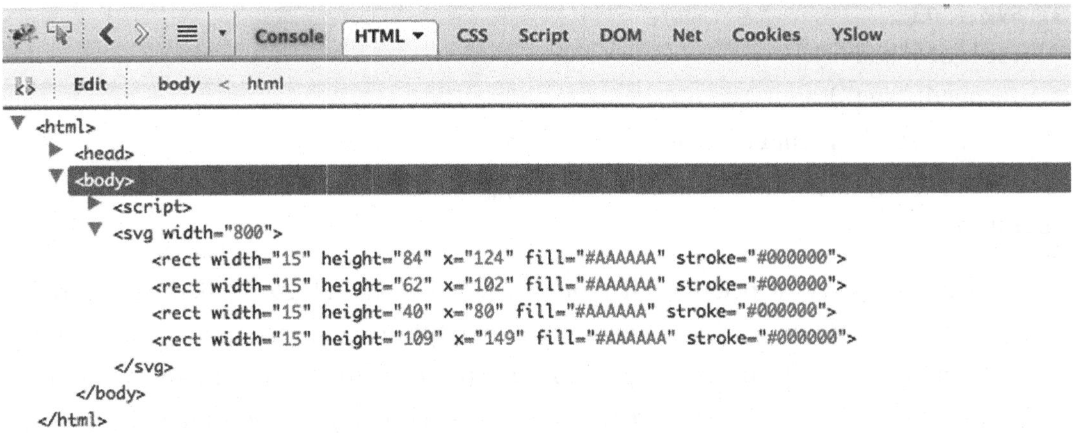

Figure 4-7. *Rectangles shown as SVG source code in Firebug*

Now you can really see the beginnings of how we can start to make data visualizations with D3 by binding data to SVG shapes. Let's take this concept another step forward.

Creating a Bar Chart

Our example so far looks a lot like the start of a bar chart in that we have a number of bars whose heights represent data. Let's give it some structure.

First, let's give our SVG container a more concrete width and height. This is important because the size of the SVG container is what determines the scale we use to normalize the rest of the chart. And because we will reference this sizing throughout our code, let's make sure we abstract these values into their own variables.

We will define a height and width for our SVG container. We'll also create variables that will hold the minimum and maximum values that we will use on our axes: 0 and 109 (the largest data point), respectively. We'll also define an offset value so we can draw the SVG container slightly larger than our chart to give the chart margins around it.

```
<script>
var chartHeight = 460,
    chartWidth = 400,
    chartMin = 0,
    chartMax = 109,
    offset = 60
```

```
var svg = d3
    .select("body")
    .append("svg")
    .attr("width", chartWidth)
    .attr("height", chartHeight + offset);
</script>
```

We next need to fix the orientation of our bars. As shown in Figure 4-6, the bars are drawn from the top down, so that although their heights are accurate, they appear to be facing down because SVG draws and positions shapes from the top left. So to get them correctly oriented so the bars look like they are coming up from the bottom of the chart, let's add a y attribute to our bars.

The y attribute should be a function that references the data; this function should subtract the bar height value from the chart height. The returned value from this function is the value used in the y coordinate.

```
<script>
bars
    .attr("width", 15 )
    .attr("height", function(x){return x;})
    .attr("y", function(x){return (chartHeight - x);})
    .attr("x", function(x){return x;})
    .attr("fill", "#AAAAAA")
    .attr("stroke", "#000000");
</script>
```

This flips the bars to the bottom of the SVG element. We can see the results in Figure 4-8.

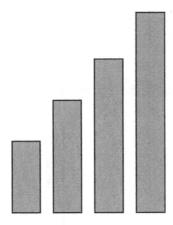

Figure 4-8. *Rectangles in bar chart no longer inverted*

Now let's scale the bars to fit the height of the SVG element. To do this, we'll use a D3 scale() function. The scale() function is used to take a number within a range and transform it to the equivalent of that number in a different range of numbers, essentially to scale values to equivalent values.

In this case, we have a number range that signifies the range of values in our dataSet array, which signify the heights of the bars, and we want to transform these numbers to equivalent values:

```
<script>
var yscale = d3.scaleLinear()
    .domain([chartMin,chartMax])
    .range([0,(chartHeight)]);
</script>
```

Be sure to place this code after the section that declares the chart variables, preferably right before we declare the "svg" variable. We then just update the height and y attributes of the bars to use the yscale() function:

```
<script>
bars
    .attr("width", 15 )
    .attr("height", function(x){ return yscale(x);})
    .attr("y", function(x){return (chartHeight - yscale(x));})
    .attr("x", function(x){return x;})
```

```
    .attr("fill", "#AAAAAA")
    .attr("stroke", "#000000");
</script>
```

This produces the graphic shown in Figure 4-9.

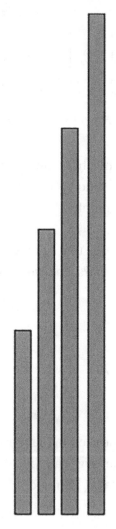

Figure 4-9. *Rectangles for bar chart properly scaled*

Very nice! But so far, we've just been placing the bars based on their height instead of where they lie in the array. Let's change that to make their array location more meaningful, so the bars are displayed in the correct order.

To do that, we just update the x value of the bars. We've seen already that we can pass in an anonymous function to the value parameter of the `attr()` function. The first parameter in our anonymous function is the value of the current element of our array. If we specify a second parameter in our anonymous function, it will hold the current index number.

We can then reference that value and offset it to place each bar:

```
<script>
bars
    .attr("width", 15 )
    .attr("height", function(x){ return yscale(x);})
    .attr("y", function(x){return (chartHeight - yscale(x));})
    .attr("x", function(x, i){return (i * 20);})
    .attr("fill", "#AAAAAA")
    .attr("stroke", "#000000");
</script>
```

This gives us the ordering of the bars shown in Figure 4-10. Just by eyeballing it, we can tell that the bars are now closer representations of the data in the array—not just the height but also the height in the order specified in the array.

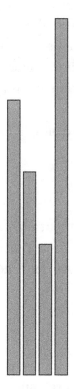

Figure 4-10. *Rectangles in bar chart ordered to follow the ordering in our data*

Now let's add text labels so that we can better see what values the heights of the bars are signifying.

We do that by creating SVG text elements in much the same way as creating the bars. We create text placeholders for every element in our data array and then style the text elements. You'll notice that the anonymous function that we pass into the x- and y-attribute calls is almost the same for the text elements as it was for the bars, only offset so that the text is above and to the center of each bar:

```
<script>
svg.selectAll("text")
    .data(dataSet)
    .enter()
    .append("text")
    .attr("x", function(d, i) { return ((i * 20) + offset/4); })
    .attr("y", function(x, i){return (chartHeight - yscale(x) - 24) ;})
    .attr("dx", -15/2)
    .attr("dy", "1.2em")
```

```
        .attr("text-anchor", "middle")
        .text(function(d) { return d;})
        .attr("fill", "black");
</script>
```

This code produces the chart shown in Figure 4-11.

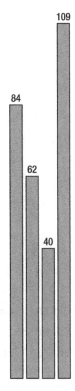

Figure 4-11. *Bar chart with text labels*

See the following complete source code:

```
<html>
<head>
<title></title>
<script src="d3.js"></script>
</head>
<body>
<script>
var dataSet = [84,62,40,109];
```

```
var chartHeight = 460,
    chartWidth = 400,
    chartMin = 0,
    chartMax = 115,
    offset = 60;
var yscale = d3.scaleLinear()
    .domain([chartMin,chartMax])
    .range([0,(chartHeight)]);
var svg = d3
    .select("body")
    .append("svg")
    .attr("width", chartWidth)
    .attr("height", chartHeight + offset);
bars = svg
    .selectAll("rect")
    .data(dataSet)
    .enter()
    .append("rect");
bars
    .attr("width", 15 )
    .attr("height", function(x){ return yscale(x);})
    .attr("y", function(x){return (chartHeight - yscale(x));})
    .attr("x", function(x, i){return (i * 20);})
    .attr("fill", "#AAAAAA")
    .attr("stroke", "#000000");
svg.selectAll("text")
    .data(dataSet)
    .enter()
    .append("text")
    .attr("x", function(d, i) { return ((i * 20) + offset/4); })
    .attr("y", function(x, i){return (chartHeight - yscale(x) - 24) ;})
    .attr("dx", -15/2)
    .attr("dy", "1.2em")
    .attr("text-anchor", "middle")
```

```
    .text(function(d) { return d;})
    .attr("fill", "black");
</script>
</body>
</html>
```

And finally, let's read in our data from external files instead of hard-coding it in the page.

Loading External Data

First, we'll take the array out of our file and put it in its own external file: sampleData.csv. The contents of sampleData.csv are simply the following:

```
84,62,40,109
```

Next, we will use the d3.text() function to load in sampleData.csv. The way d3.text() works is that it takes a path to an external file and then assigns it to a variable (in this case named data). The function receives a parameter that is the contents of the external file:

```
<script>
d3.text("sampleData.csv").then((data) => {});
</script>
```

The catch is that we need the contents of our external file before we can begin doing any charting on the data. So within the callback function, we will parse up the file and then wrap all our existing functionality, like so:

```
<html>
<head>
<title></title>
<script src="d3.js"></script>
</head>
<body>
<script>
d3.text("sampleData.csv").then((data) => {
var dataSet = data.split(",");
```

```
var chartHeight = 460,
    chartWidth = 400,
    chartMin = 0,
    chartMax = 115,
    offset = 60;
var yscale = d3.scaleLinear()
    .domain([chartMin,chartMax])
    .range([0,(chartHeight)]);
var svg = d3
    .select("body")
    .append("svg")
    .attr("width", chartWidth)
    .attr("height", chartHeight + offset);
bars = svg
    .selectAll("rect")
    .data(dataSet)
    .enter()
    .append("rect");
bars
    .attr("width", 15 )
    .attr("height", function(x){ return yscale(x);})
    .attr("y", function(x){return (chartHeight - yscale(x));})
    .attr("x", function(x, i){return (i * 20);})
    .attr("fill", "#AAAAAA")
    .attr("stroke", "#000000");
svg.selectAll("text")
    .data(dataSet)
    .enter()
    .append("text")
    .attr("x", function(d, i) { return ((i * 20) + offset/4); })
    .attr("y", function(x, i){return (chartHeight - yscale(x) - 24) ;})
    .attr("dx", -15/2)
    .attr("dy", "1.2em")
    .attr("text-anchor", "middle")
```

```
    .text(function(d) { return d;})
    .attr("fill", "black");
  })
</script>
</body>
</html>
```

It's important to note that if you are running this code locally on your computer, as opposed to on a web server, you will get an error similar to "Cross origin requests are only supported for HTTP." This is a security measure that your browser is using in order to prevent malicious code from running on your local machine. It's advised to use a local web server to work around this issue while programming.

Returning to our d3.text() function—CSV files aren't the only format we can read in. In fact, `d3.text()` is only syntactic sugar—a convenience method or a type-specific wrapper for D3's implementation of the XMLHttpRequest object `d3.xhr()`.

For reference, the XMLHttpRequest object is what is used in AJAX transactions to load content asynchronously from the client side without refreshing the page. In pure JavaScript, we instantiate the XHR object, pass in a URL to a resource, and the method to retrieve the resource (GET or POST). We also specify a callback function that will get invoked when the XHR object is updated. In this function, we can parse up the data and begin using it. See Figure 4-12 for a high-level diagram of this process.

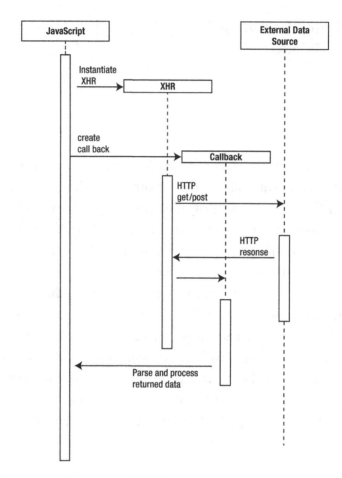

Figure 4-12. *Sequence diagram of XHR transaction*

In D3, the d3.xhr() function is D3's wrapper for the XMLHttpRequest object. It works much the same way that we just saw d3.text() work, where we pass in a URL to a resource and a callback function to execute.

The other type-specific convenience functions that D3 has are d3.csv(), d3.json(), d3.xml(), and d3.html().

Summary

This chapter explored D3. We started out covering the introductory concepts of HTML, CSS, SVG, and JavaScript, at least the points that are pertinent to implementing D3. From there, we delved into D3, looking at introductory concepts like creating our first SVG shapes to expanding on that idea by making those shapes into a bar graph.

D3 is a fantastic library for crafting data visualizations. To see the full API documentation, see `https://github.com/mbostock/d3/wiki/API-Reference`.

We will return to D3, but first, we will explore some data visualizations that we can create that have practical application in the world of web development. The first one we will look at is something that you may have seen in your Google analytics dashboard or something similar: a data map based on user visits.

CHAPTER 5

Visualizing Spatial Data from Access Logs

In the last chapter, we talked about D3 and looked at concepts from making simple shapes to creating a bar chart out of those shapes. In the previous two chapters, we took a deep dive into R. Now that you are familiar with the core technologies that we will be using, let's begin looking at examples of how, as web developers, we can create data visualizations that communicate useful information around our domain.

The first one that we will look at is creating a data map out of our access logs.

What Are Data Maps?

First, let's level set and make sure that we clearly define a data map. A data map is a representation of information over a spatial field, a marriage of statistics with cartography. Data maps are some of the most easily understood and widely used data visualizations there are because their data is couched in something that we are all familiar with and use anyway: maps.

Recall the discussion in Chapter 1 of the Cholera map created by Jon Snow in 1854. This is considered one of the earliest examples of a data map, though there are several notable contemporaries, including several by Charles Minard, an engineer in nineteenth-century France. He is most widely remembered for his data visualization of Napoleon's invasion of Russia in 1812.

© Tom Barker, Jon Westfall 2022
T. Barker and J. Westfall, *Pro Data Visualization Using R and JavaScript*,
https://doi.org/10.1007/978-1-4842-7202-2_5

Minard also created several prominent data maps. Two of his most famous data maps include the data map demonstrating the source region and percentage of total cattle consumed in France (see Figure 5-1) and the data map demonstrating the wine export path and destination from France (see Figure 5-2).

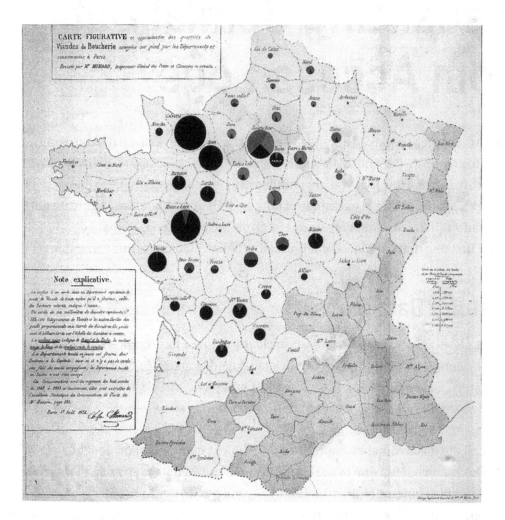

Figure 5-1. *Early data map from Charles Minard demonstrating source region and cattle consumption in France*

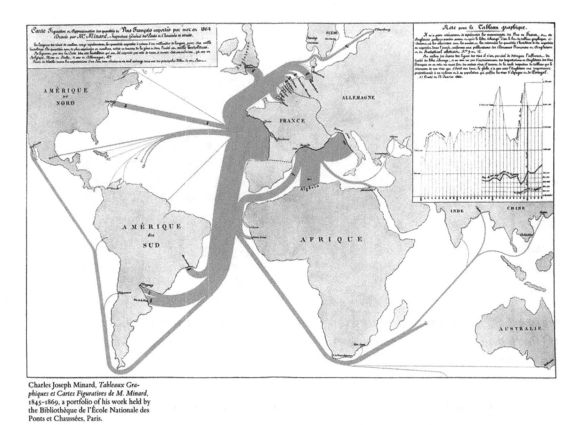

Charles Joseph Minard, *Tableaux Gra-
phiques et Cartes Figuratives de M. Minard*,
1845–1869, a portfolio of his work held by
the Bibliothèque de l'École Nationale des
Ponts et Chaussées, Paris.

Figure 5-2. *Data map from Minard demonstrating wine export path and
destination*

Today, we see data maps everywhere. They can be informative and artistic
expressions, like the wind map project from Fernanda Viegas and Martin Wattenberg
(see Figure 5-3). Available at `http://hint.fm/wind`, the wind project demonstrates the
path and force of wind currents over the United States.

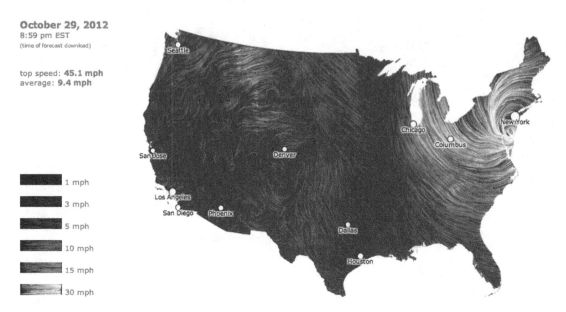

October 29, 2012
8:59 pm EST
(time of forecast download)

top speed: **45.1 mph**
average: **9.4 mph**

1 mph
3 mph
5 mph
10 mph
15 mph
30 mph

Figure 5-3. *Wind map, showing wind speeds by region for the touchdown of Hurricane Sandy (used with permission of Fernanda Viegas and Martin Wattenberg)*

Data maps can be profound, such as those available at energy.gov that demonstrate concepts such as energy consumption by state (see Figure 5-4) or even renewable energy production by state.

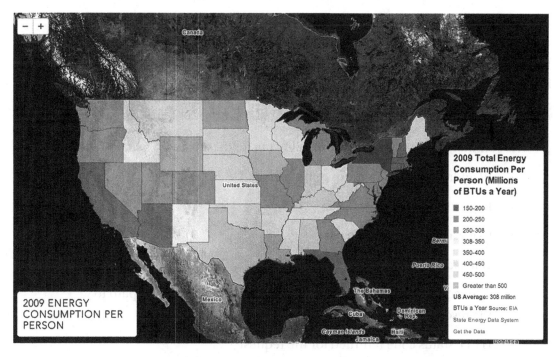

Figure 5-4. *Data map depicting energy consumption by state, from energy.gov (available at* `http://energy.gov/maps/2009-energy-consumption-person`*)*

You've now seen historical and contemporary examples of data maps. In this chapter, you will look at creating your own data map from web server access logs.

Access Logs

Access logs are records that a web server keeps to track what resources were requested. Whenever a web page, an image, or any other kind of file is requested from a server, the server makes a log entry for the request. Each request has certain data points associated with it, usually information about the requestor of the resource (e.g., IP address and user agent) and general information such as time of day and what resource was requested.

Let's look at an access log. A sample entry looks like this:

```
msnbot-157-55-17-199.search.msn.com - - [18/Jan/2013:13:32:15 -0400] "GET
/robots.txt HTTP/1.1" 404 208 "-" "Mozilla/5.0 (compatible; bingbot/2.0;
+ http://www.bing.com/bingbot.htm)"
```

121

This is a snippet from a sample Apache access log. Apache access logs follow the combined log format, which is an extension of the common log format standard of the World Wide Web Consortium (W3C). Documentation for the common log format can be found here:

`www.w3.org/Daemon/User/Config/Logging.html#common-logfile-format`

The common log format defines the following fields, separated by tabs:

- IP address or DNS name of remote host

- Logname of the remote user

- Username of the remote user

- Datestamp

- The request—usually includes the request method and the path to the resource requested

- HTTP status code returned for the request

- Total file size of the resource requested

The combined log format adds the referrer and user agent fields. The Apache documentation for the combined log format can be found here:

`http://httpd.apache.org/docs/current/logs.html#combined`

Note that fields that are not available are represented by a single dash -. Let's dissect the previous log entry:

- The first field is `msnbot-157-55-17-199.search.msn.com`. This is a DNS name that just happens to have the IP address built into it. We can't count on parsing the IP address out of this domain, so for now, just ignore the IP address. When we get to programmatically parsing the logs, we will use the native PHP function `gethostbyname()` to look up the IP addresses for given domain names.

- The next two fields, the logname and the user, are empty.

- Next is the datestamp: `[18/Jan/2013:13:32:15 -0400]`.

- After the datestamp is the request: `"GET /robots.txt HTTP/1.1"`. If you hadn't already guessed from the DNS name, this is a bot, specifically Microsoft's `msnbot` replacement: the `bingbot`. In this record, the `bingbot` is requesting the `robots.txt` file.

- Next is the HTTP status of the request: `404`. Clearly, there was no `robots.txt` file available.

- Next is the total payload of the request. Apparently the 404 cost 208 bytes.

- Next is a dash to signify that the referrer was empty.

- The last is the useragent: `"Mozilla/5.0 (compatible; bingbot/2.0; +http://www.bing.com/bingbot.htm)"`, which tells us definitively that it is indeed a bot.

Now that you have the access log and understand what is in it, you can parse it to use each field in it programmatically.

Parsing the Access Log

The process of parsing the access log is the following:

1. Read in the access log.

2. Parse it and gather geographic data based on the stored IP address.

3. Output the fields that we are interested in for our visualization.

4. Read in this output and visualize.

We'll use PHP for the first three steps and R for the last step. Note that you will need to be running PHP 5.4.10 or higher to successfully run the following PHP code.

Read in the Access Log

Create a new PHP document called `parseLogs.php`, within which you will first create a function to read in a file. Call this function `parseLog()` and have it accept the path to the file:

```
function parseLog($file){
}
```

Within this function, you will write some code that will open the passed-in file for reading and iterate through each line of the file until it reaches the end of the file. Each step in the iteration stores the line that is read in, in the variable $line:

```
$logArray = array();
$file_handle = fopen($file, "r");
while (!feof($file_handle)) {
    $line = fgets($file_handle);
}
fclose($file_handle);
```

Fairly standard file I/O functionality in PHP so far. Within the loop, you will stub out a function call to a function that you will call `parseLogLine()` and another function that you will call `getLocationbyIP()`. In `parseLogLine()`, you will split up the line and store the values in an array. In `getLocationbyIP()`, you will use the IP address to get geographic information. You will then store this returned array in a larger array that is called $logArray.

```
$lineArr = parseLogLine($line);
$lineArr = getLocationbyIP($lineArr);
$logArray[count($logArray)] = $lineArr;
```

Don't forget to create the $logArray variable at the top of the function.
The finished function should look like so:

```
function parseLog($file){
$logArray = array();
$file_handle = fopen($file, "r");
while (!feof($file_handle)) {
    $line = fgets($file_handle);
```

```
    $lineArr = parseLogLine($line);
    $lineArr = getLocationbyIP($lineArr);
    $logArray[count($logArray)] = $lineArr;
}
fclose($file_handle);
return $logArray;
}
```

Parse the Log File

Next, you'll flesh out the parseLogLine() function. First, you'll create the empty function:

```
function parseLogLine($logLine){
}
```

The function will expect a single line of the access log.

Remember that each line of the access log is made up of sections of information separated by whitespace. Your first instinct might be to just split the line at each instance of a whitespace, but this would result in breaking up the user agent string (and potentially other fields) in unexpected ways.

For our purposes, a much cleaner way to parse the line is to use a regular expression. Regular expressions, called regex for short, are patterns that enable you to do quick and efficient string matching.

Regular expressions use special characters to define these patterns: individual characters, character literals, or sets of characters. A deep dive on regular expressions is outside of the scope of this chapter, but a great reference to read about the different regular expression patterns is the Microsoft regular expression Quick Reference, available here: http://msdn.microsoft.com/en-us/library/az24scfc.aspx.

Grant Skinner also provides a great tool for creating and debugging regular expressions (see Figure 5-5), which is available here: https://regexr.com.

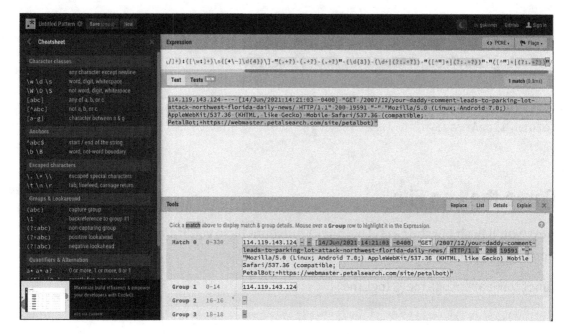

Figure 5-5. *Grant Skinner's regex tool*

To use Grant's tool, change the mode at the top from JavaScript to PCRE (which is how PHP interprets regular expressions). Then paste in the following into the large "Text" box:

114.119.143.124 - - [14/Jun/2021:14:21:03 -0400] "GET /2007/12/your-daddy-comment-leads-to-parking-lot-attack-northwest-florida-daily-news/ HTTP/1.1" 200 19591 "-" "Mozilla/5.0 (Linux; Android 7.0;) AppleWebKit/537.36 (KHTML, like Gecko) Mobile Safari/537.36 (compatible; PetalBot;+https://webmaster.petalsearch.com/site/petalbot)"

Finally, enter the following regular expression into the "Expression" box: ^([\d.:]+) (\S+) (\S+) \[([\w\/]+):([\w:]+)\s([+\-]\d{4})\] "(.+?) (.+?) (.+?)" (\d{3}) (\d+|(?:.+?)) "([^"]*|(?:.+?))" "([^"]*|(?:.+?))"

Clicking the expression match will now let you explore how each portion of the regular expression is found in the log entry that we pasted in.

Turning to our PHP code, let's define our regular expression pattern and store it in a variable that we will call $pattern.

If you aren't proficient with regex, you can create them fairly easily using Grant Skinner's tool (refer to Figure 5-5). Using this tool, you can come up with the following pattern:

```
$pattern = '/^([\d.:]+) (\S+) (\S+) \[([\w\/]+):([\w:]+)\s([+\-
]\d{4})\] "(.+?) (.+?) (.+?)" (\d{3}) (\d+|(?:.+?)) "([^"]*|(?:.+?))"
"([^"]*|(?:.+?))"/';
```

Within the tool, you can see how it breaks up the strings into the following groups (see Figure 5-6).

Group 1	0–14	114.119.143.124
Group 2	16–16	-
Group 3	18–18	-
Group 4	21–31	14/Jun/2021
Group 5	33–40	14:21:03
Group 6	42–46	-0400
Group 7	50–52	GET
Group 8	54–138	/2007/12/your-daddy-comment-leads-to-parking-lot-attack-northwest-florida-daily-news/

Figure 5-6. *Log file line split into groups*

You now have a regular expression to use. Let's use PHP's `preg_match()` function. This takes as parameters a regular expression, a string to match it against, and an array to populate as the output of the pattern matching:

```
preg_match($pattern,$logLine,$logs);
```

From there, we can just create an associative array with named indexes to hold our parsed up line:

```
$logArray = array();
$logArray['ip'] = gethostbyname($logs[1]);
$logArray['identity'] = $logs[2];
$logArray['user'] = $logs[2];
$logArray['date'] = $logs[4];
$logArray['time'] = $logs[5];
$logArray['timezone'] = $logs[6];
$logArray['method'] = $logs[7];
```

```
$logArray['path'] = $logs[8];
$logArray['protocol'] = $logs[9];
$logArray['status'] = $logs[10];
$logArray['bytes'] = $logs[11];
$logArray['referer'] = $logs[12];
$logArray['useragent'] = $logs[13];
```

Our complete parseLogLine() function should now look like this:

```
function parseLogLine($logLine){
        $pattern = '/^([\d.:]+) (\S+) (\S+) \[(([\w\/]+):([\w:]+)\
        s([+\-]\d{4}))\] "(.+?) (.+?) (.+?)" (\d{3}) (\d+|(?:.+?))
        "([^"]*|(?:.+?))" "([^"]*|(?:.+?))"/';
        preg_match($pattern,$logLine,$logs);
        $logArray = array();
        $logArray['ip'] = gethostbyname($logs[1]);
        $logArray['identity'] = $logs[2];
        $logArray['user'] = $logs[2];
        $logArray['date'] = $logs[4];
        $logArray['time'] = $logs[5];
        $logArray['timezone'] = $logs[6];
        $logArray['method'] = $logs[7];
        $logArray['path'] = $logs[8];
        $logArray['protocol'] = $logs[9];
        $logArray['status'] = $logs[10];
        $logArray['bytes'] = $logs[11];
        $logArray['referer'] = $logs[12];
        $logArray['useragent'] = $logs[13];
        return $logArray;
}
```

Next, you will create the functionality for the getLocationbyIP() function.

Geolocation by IP

In the getLocationbyIP() function, you can take the array that you made by parsing a line of the access log and use the IP field to get the geographic location. There are many ways to get geographic location by IP address; most involve either calling a third-party API or downloading a third-party database with the IP location information prepopulated. Some of these third parties are freely available; some have a cost associated with them.

For our purposes, you can use the free API available at hostip.info. Figure 5-7 shows the hostip.info home page.

Figure 5-7. hostip.info home page

The hostip.info service aggregates geotargeting information from ISPs as well as direct feedback from users. It exposes an API as well as a database available for download.

The API is available at `http://api.hostip.info/`. If no parameters are provided, the API returns the geolocation of the client. By default, the API returns XML. The return value looks like this:

```xml
<?xml version="1.0" encoding="ISO-8859-1" ?>
<HostipLookupResultSet version="1.0.1" xmlns:gml=" http://www.opengis.net/
gml " xmlns:xsi=" http://www.w3.org/2001/XMLSchema-instance " xsi:noName
spaceSchemaLocation=" http://www.hostip.info/api/hostip-1.0.1.xsd ">
 <gml:description>This is the Hostip Lookup Service</gml:description>
 <gml:name>hostip</gml:name>
 <gml:boundedBy>
  <gml:Null>inapplicable</gml:Null>
 </gml:boundedBy>
 <gml:featureMember>
  <Hostip>
   <ip>71.225.152.145</ip>
   <gml:name>Chalfont, PA</gml:name>
   <countryName>UNITED STATES</countryName>
   <countryAbbrev>US</countryAbbrev>
   <!-- Co-ordinates are available as lng,lat -->
   <ipLocation>
    <gml:pointProperty>
     <gml:Point srsName=" http://www.opengis.net/gml/srs/epsg.xml#4326 ">
      <gml:coordinates>-75.2097,40.2889</gml:coordinates>
     </gml:Point>
    </gml:pointProperty>
   </ipLocation>
  </Hostip>
 </gml:featureMember>
</HostipLookupResultSet>
```

You can refine the API calls. If you want only country information, you can call `http://api.hostip.info/country.php`. It returns a string with a country code. If JSON is preferred over XML, you can call `http://api.hostip.info/get_json.php` and get the following result:

```
{"country_name":"UNITED STATES","country_code":"US","city":"Chalfont,
PA","ip":"71.225.152.145"}
```

To specify an IP address, add the parameter `?ip=xxxx`, like so:

```
http://api.hostip.info/get_json.php?ip=100.43.83.146
```

OK, let's code the function!

We'll stub out the function and have it accept an array. We'll pull the IP address from the array, store it in a variable, and concatenate the variable to a string that contains the path to the hostip.info API:

```
function getLocationbyIP($arr){
    $IPAddress = $arr['ip'];
    $IPCheckURL = " http://api.hostip.info/get_json.php?ip=$IPAddress ";
}
```

You'll pass this string to the native PHP function `file_get_contents()` and store the return value, the results of the API call, in a variable that you'll name `jsonResponse`. You'll use the PHP `json_decode()` function to convert the returned JSON data into a native PHP object:

```
$jsonResponse = file_get_contents($IPCheckURL);
$geoInfo = json_decode($jsonResponse);
```

You next pull the geolocation data from the object and add it to the array that you passed into the function. The city and state information is a single string separated by a comma and a space ("Philadelphia, PA"), so you'll need to split at the comma and save each field separately in the array.

```
$arr['country'] = $geoInfo->{"country_code"};
$arr['city'] = explode(",",$geoInfo->{"city"})[0];
$arr['state'] = explode(",",$geoInfo->{"city"})[1];
```

Next, let's do a little bit of error checking that will make things easier later on in the process. You'll check to see whether the state string has any value; if it doesn't, set it to "XX". This will be helpful once you begin parsing data in R. And finally, you'll return the updated array:

```
if(count($arr['state']) < 1)
    $arr['state'] = "XX";
return $arr;
```

The full function should look like this:

```
function getLocationbyIP($arr){
    $IPAddress = $arr['ip'];
    $IPCheckURL = " http://api.hostip.info/get_json.php?ip=$IPAddress ";
    $jsonResponse = file_get_contents($IPCheckURL);
    $geoInfo = json_decode($jsonResponse);
    $arr['country'] = $geoInfo->{"country_code"};
    $arr['city'] = explode(",",$geoInfo->{"city"})[0];
    $arr['state'] = explode(",",$geoInfo->{"city"})[1];
    if(count($arr['state']) < 1)
        $arr['state'] = "XX";
    return $arr;
}
```

Finally, let's create a function to write processed data out to a file.

Output the Fields

You'll create a function named writeRLog() that accepts two parameters—the array populated with decorated log data and the path to a file:

```
function writeRLog($arr, $file){
}
```

You need to create a variable called writeFlag that will be the flag to tell PHP to either write or append data to the file. You check to see whether the file exists; if it does, you append content instead of overwrite. After the check, open the file:

```
writeFlag = "w";
if(file_exists($file)){
    $writeFlag = "a";
}
$fh = fopen($file, $writeFlag) or die("can't open file");
```

You then loop through the passed-in array; construct a string containing the IP address, date, HTTP status, country code, state, and city of each log entry; and write that string to the file. Once you've finished iterating through the array, you close the file.

```
for($x = 0; $x < count($arr); $x++){
    if($arr[$x]['country'] != "XX"){
        $data = $arr[$x]['ip'] . "," . $arr[$x]['date'] . "," . $arr[$x]
        ['status'] . "," . $arr[$x]['country'] . "," . $arr[$x]['state']
        . "," . $arr[$x]['city'];
    }
    fwrite($fh, $data . "\n");
 }
```

Our completed writeRLog() function should look like this:

```
function writeRLog($arr, $file){
    $writeFlag = "w";
    if(file_exists($file)){
        $writeFlag = "a";
        }
        $fh = fopen($file, $writeFlag) or die("can't open file");
        for($x = 0; $x < count($arr); $x++){
            if($arr[$x]['country'] != "XX"){
                $data = $arr[$x]['ip'] . "," . $arr[$x]['date'] . "," .
                $arr[$x]['status'] . "," . $arr[$x]['country'] . "," .
                $arr[$x]['state'] . "," . $arr[$x]['city'];
            }
            fwrite($fh, $data . "\n");
```

```
        }
    fclose($fh);
    echo "log created";
}
```

Adding Control Logic

Finally, you'll create some control logic to invoke all these functions that you just created. You'll declare the path to the access log and the path to our output flat file, call parseLog(), and send the output to writeRLog().

```
$logfile = "access_log";
$chartingData = "accessLogData.txt";
$logArr = parseLog($logfile);
writeRLog($logArr, $chartingData);
```

Our completed PHP code should look like the following:

```
<html>
<head></head>
<body>
<?php
$logfile = "access_log";
$chartingData = "accessLogData.txt";
$logArr = parseLog($logfile);
writeRLog($logArr, $chartingData);
function parseLog($file){
        $logArray = array();
        $file_handle = fopen($file, "r");
        while (!feof($file_handle)) {
            $line = fgets($file_handle);
            $lineArr = parseLogLine($line);
            $lineArr = getLocationbyIP($lineArr);
            $logArray[count($logArray)] = $lineArr;
        }
```

```php
        fclose($file_handle);
        return $logArray;
}
function parseLogLine($logLine){
        $pattern = '/^([\d.:]+) (\S+) (\S+) \[([\w\/]+):([\w:]+)\
        s([+\-]\d{4})\] "(.+?) (.+?) (.+?)" (\d{3}) (\d+|(?:.+?))
        "([^"]*|(?:.+?))" "([^"]*|(?:.+?))"/';
        preg_match($pattern,$logLine,$logs);
        $logArray = array();
        $logArray['ip'] = gethostbyname($logs[1]);
        $logArray['identity'] = $logs[2];
        $logArray['user'] = $logs[2];
        $logArray['date'] = $logs[4];
        $logArray['time'] = $logs[5];
        $logArray['timezone'] = $logs[6];
        $logArray['method'] = $logs[7];
        $logArray['path'] = $logs[8];
        $logArray['protocol'] = $logs[9];
        $logArray['status'] = $logs[10];
        $logArray['bytes'] = $logs[11];
        $logArray['referer'] = $logs[12];
        $logArray['useragent'] = $logs[13];
        return $logArray;
}
function getLocationbyIP($arr){
        $IPAddress = $arr['ip'];
        $IPCheckURL = "http://api.hostip.info/get_json.php?ip=$IPAddress";
        $jsonResponse =  file_get_contents($IPCheckURL);
        $geoInfo = json_decode($jsonResponse);
        $arr['country'] = $geoInfo->{"country_code"};
        $arr['city'] = explode(",",$geoInfo->{"city"})[0];
        $arr['state'] = explode(",",$geoInfo->{"city"})[1];
        return $arr;
}
```

```
function writeRLog($arr, $file){
        $writeFlag = "w";
        if(file_exists($file)){
                $writeFlag = "a";
        }
        $fh = fopen($file, $writeFlag) or die("can't open file");
        for($x = 0; $x < count($arr); $x++){
                if($arr[$x]['country'] != "XX"){
                        $data = $arr[$x]['ip'] . "," . $arr[$x]['date']
                        . "," . $arr[$x]['status'] . "," . $arr[$x]
                        ['country'] . "," . $arr[$x]['state'] . "," .
                        $arr[$x]['city'];
                }
                fwrite($fh, $data . "\n");
        }
        fclose($fh);
        echo "log created";
}
?>
</body>
</html>
```

And it should produce a flat file that looks similar to this:

```
71.225.152.145,18/Jan/2013,404,US, PA,Chalfont
114.119.143.124,14/Jun/2021,200,AU,,Canberra
```

We have made a sample access log available here: https://jonwestfall.com/data/access_log.

Creating a Data Map in R

So far, you parsed the access log, scrubbed the data, decorated it with location information, and created a flat file that has a subset of information. The next step is to visualize this data.

Because you are making a map, you need to install the map package. Open up R; from the console, type the following:

```
> install.packages('maps')
> install.packages('mapproj')
```

Now we can begin! To reference the map package in the R script, you need to load it into memory by calling the library() function:

```
library(maps)
library(mapproj)
```

You next create several variables—one to point to our formatted access log data; another is a list of column names. You create a third variable, logData, to hold the data frame created when you read in the flat file.

```
logDataFile <- '/Applications/MAMP/htdocs/accessLogData.txt'
logColumns <- c("IP", "date", "HTTPstatus", "country", "state", "city")
logData <- read.table(logDataFile, sep=",", col.names=logColumns)
```

If you type **logData** in the console, you see the data frame formatted like this:

```
> logData
    IP              date          HTTPstatus   country   state   city
1   100.43.83.146   25/Jan/2013   404          US        NV      Las Vegas
2   100.43.83.146   25/Jan/2013   301          US        NV      Las Vegas
3   64.29.151.221   25/Jan/2013   200          US        XX      (Unknown city)
4   180.76.6.26     25/Jan/2013   200          CN        XX      Beijing
```

Clearly, you could start to track several different data points here. Let's first look at mapping out what countries the traffic is coming from.

Mapping Geographic Data

You can begin by pulling the unique country names from logData. You'll store this in a variable named country:

```
> country <- unique(logData$country)
```

137

If you type **country** in the console, the data looks like the following:

```
> country
[1] US CN CA SE UA
Levels: CA CN SE UA US
```

These are the country codes that you get back from iphost.info. R has a different set of country codes that it uses, so you'll need to convert the iphost country codes to R country codes. You can do this by applying a function to the country list.

You'll use `sapply()` to apply an anonymous function of your own design to the list of country codes. In the anonymous function, you'll trim any whitespace and do a direct replacement of country codes. You will use the `gsub()` function to do a replacement of all instances of the passed-in parameter.

```
country <- sapply(country, function(countryCode){
  #trim whitespaces from the country code
  countryCode <- gsub("(^ +)|( +$)", "", countryCode)
  if(countryCode == "US"){
    countryCode<- "USA"
  }else if(countryCode == "AU"){
    countryCode<- "Australia"
  }}
)
```

You'll notice that you are hard-coding every country code that you have. This is, of course, bad form, and you'll approach this problem a very different way once you dig into state data.

If you type **country** into the console again, you'll now see the following:

```
> country
        US            AU
     "USA" "Australia"
```

You next use the `match.map()` function to match the countries with the map package's list of countries. The `match.map()` function creates a numeric vector in which each element corresponds to a country on the world map. The elements of intersection (where countries in the country list match countries in the world map) have values assigned to them—specifically, the index number from the original country list. So the element that corresponds to USA has a 1, the element that corresponds to Canada has a 2, and so on. Where there is no intersection, the element has the value NA.

```
countryMatch <- match.map("world2", country)
```

Let's next use the `countryMatch` list to create a color-coded country match. To do this, simply apply a function that checks each element. If it is not NA, assign the color #C6DBEF to the element, which is a nice light blue. If the element is NA, set the element to white or #FFFFFF. You will save the result of this in a new list that you will call `colorCountry`.

```
colorCountry <- sapply(countryMatch, function(c){
 if(!is.na(c)) c <- "#C6DBEF"
 else c <- "#FFFFFF"
})
```

Now let's create our first visualization with the `map()` function! The `map()` function accepts several parameters:

- The first is the name of the database to use. The database name can be either `world`, `usa state`, or `county`; each contains data points that correlate to geographic areas that the `map()` function will draw.

- If you only want to draw a subset of the larger geographic database, you can specify an optional parameter named `region` that lists the areas to draw.

- You can also specify the map projection to use. A *map projection* is basically a way to represent a three-dimensional curved space on a flat surface. There are a number of predefined projections, and the `mapproj` package in R supports a number of these. For the world map that you'll be making, you will use an equal area projection, the identifier of which is "azequalarea". For more about map projections, see `http://xkcd.com/977/`.

- You also can specify the center point of our map, in latitude and longitude, using the `orientation` parameter.

- Finally, you'll pass the `colorCountry` list that you just made to the `col` parameter.

```
map('world', proj='azequalarea', orient=c(41,-74,0), boundary=TRUE,
col=colorCountry, fill=TRUE)
```

This code produces the map that you can see in Figure 5-8.

Figure 5-8. *Data map using a world map*

From this map, we can see that the countries from our unique list are shaded blue and the rest of the countries are colored white. This is good, but we can make it better.

Adding Latitude and Longitude

Let's start by adding latitude and longitude lines, which will accentuate the curvature of the globe and give context to where the poles are. To create latitude and longitude lines, we first create a new map object, but we will set plot to FALSE so that the map is not drawn to the screen. We'll save this map object to a variable named m:

```
m <- map('world',plot=FALSE)
```

We'll next call map.grid() and pass in our stored map object:

```
map.grid(m, col="blue", label=FALSE, lty=2, pretty=TRUE)
```

Note that if you are running this code line by line in the command window, it's important to keep the Quartz graphic window open as you type the lines in so that R can update that chart. If you close the Quartz window while typing it in line by line, you could get an error stating that plot.new has not been called. Or you could type each line into a text file and copy them into the R command line all at once.

While we're at it, let's add a scale to the chart to show

```
map.scale()
```

Our completed R code should now look like so:

```
library(maps)
library(mapproj)
logDataFile <- '/Applications/MAMP/htdocs/accessLogData.txt'
logColumns <- c("IP", "date", "HTTPstatus", "country", "state", "city")
logData <- read.table(logDataFile, sep=",", col.names=logColumns)
country <- unique(logData$country)
country <- sapply(country, function(countryCode){
  #trim whitespaces from the country code
  countryCode <- gsub("(^ +)|( +$)", "", countryCode)
  if(countryCode == "US"){
    countryCode<- "USA"
  }else if(countryCode == "CN"){
    countryCode<- "China"
  }else if(countryCode == "CA"){
    countryCode<- "Canada"
```

```
  }else if(countryCode == "SE"){
    countryCode<- "Sweden"
  }else if(countryCode == "UA"){
    countryCode<- "USSR"
  }
})
countryMatch <-  match.map("world", country)
#color code any states with visit data as light blue
colorCountry <- sapply(countryMatch, function(c){
 if(!is.na(c)) c <- "#C6DBEF"
 else c <- "#FFFFFF"
})
m <- map('world',plot=FALSE)
map('world',proj='azequalarea',orient=c(41,-74,0), boundary=TRUE,
col=colorCountry,fill=TRUE)
map.grid(m,col="blue", label=FALSE, lty=2, pretty=TRUE)
map.scale()
```

And this code outputs the world map shown in Figure 5-9.

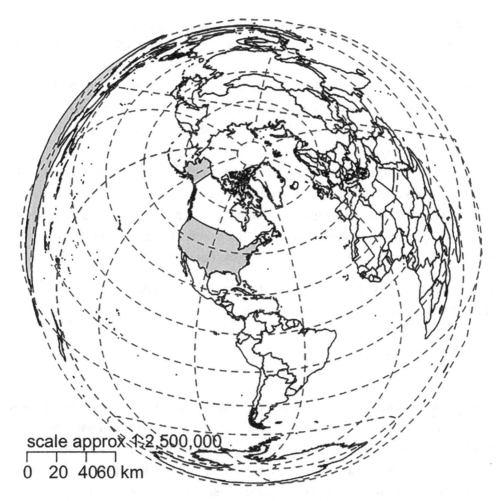

Figure 5-9. *Globe data map with latitude and longitude lines as well as scale*

Very nice! Next, let's drill into a breakdown of visits by states in the United States.

Displaying Regional Data

Let's start by isolating US data; we can do this by selecting all rows in which the state does not equal "XX". Remember setting the value in the state column to "XX" when we were parsing the access log in PHP? This is why. Countries other than the United States don't have state data associated with them, so we can simply pull only the rows that have state data.

```
usData <- logData[logData$state != "XX", ]
```

We next need to replace the state abbreviations that we got from hostip.info with the full state names so that we can create a `match.map` lookup list, much like we did with the preceding country data.

The upside with state data is that R has a data set that contains all 50 US state names, abbreviations, and even more esoteric information such as area of the state and named divisions (New England, Middle Atlantic, and so on). For more information, type **?state.name** at the R console.

We can use the information in this data set to match the state abbreviations with the full state names that the map package needs. To do this, we use the `apply()` function to run an anonymous function that greps through the `state.abb` data set to find a match for the passed-in state abbreviation and then use that returned value as the index for retrieving the full state name from the `state.name` data set:

```
usData$state <- apply(as.matrix(usData$state), 1, function(s){
  #trim the abbreviation of whitespaces
  s <- gsub("(^ +)|( +$)", "", s)
  s <- state.name[grep(s, state.abb)]
})
```

We achieve the same functionality as the previous country match, but much more elegantly. If we were so inclined, we could go back and create our own data set of country names for future use to have a similar elegant solution for the country match.

Now that we have full state names to use, we can pull a unique list of state names and use that list to create a map matched list (again, just as we did for countries):

```
states <- unique(usData$state)
stateMatch <- match.map("state", states)
```

With our state match list, we can again apply a function to it that will look for matches in our match list, elements that do not have the value NA, and set the value for those elements to our nice light blue color while all elements that do have the value of NA get set to white. We save this list in a variable that we name `colorMatch`.

```
#color code any states with visit data as light blue
colorMatch <- sapply(stateMatch, function(s){
 if(!is.na(s)) s <- "#C6DBEF"
 else s <- "#FFFFFF"
})
```

We can then use colorMatch in our call to the map() function:

```
map("state", resolution = 0,lty = 0,projection = "azequalarea",
col=colorMatch,fill=TRUE)
```

Hmm, but notice something? Only the colored areas are drawn to the stage, as shown in Figure 5-10.

Figure 5-10. *Data map with only states that have data displayed*

We need to make a second map() call that will draw the remainder of the map. In this map() call, we will set the add parameter to TRUE, which will cause the new map that we are drawing to be added to the current map. While we're at it, let's create a scale for this map as well:

```
map("state", col = "black", fill=FALSE, add=TRUE, lty=1, lwd=1,
projection="azequalarea")
map.scale()
```

This code produces the finished state map in Figure 5-11.

scale approx 1:310,000

0   5   10   15 km

Figure 5-11. *Completed state data map*

Distributing the Visualization

OK, now let's put our R code in an R Markdown file for distribution. Let's go into RStudio and click File ➤ New ➤ R Markdown. Let's add a header and make sure that our R code is wrapped in ```` ```{r} ```` tags and that our charts have heights and widths assigned to them. Our completed R Markdown file should look like this:

```
Visualizing Spatial Data from Access Logs
==========================================================
```{r}
library(maps)
library(mapproj)
logDataFile <- '/Applications/MAMP/htdocs/accessLogData.txt'
logColumns <- c("IP", "date", "HTTPstatus", "country", "state", "city")
logData <- read.table(logDataFile, sep=",", col.names=logColumns)
```

```{r fig.width=15, fig.height=10}
#chart worldwide visit data
#unfortunately there is no state.name equivalent for countries so we must check
```
```

```
#the explicit country names. In the us states below we are able to
accomplish this much
#more efficiently
country <- unique(logData$country)
country <- sapply(country, function(countryCode){
  #trim whitespaces from the country code
  countryCode <- gsub("(^ +)|( +$)", "", countryCode)
  if(countryCode == "US"){
    countryCode<- "USA"
  }else if(countryCode == "CN"){
    countryCode<- "China"
  }else if(countryCode == "CA"){
    countryCode<- "Canada"
  }else if(countryCode == "SE"){
    countryCode<- "Sweden"
  }else if(countryCode == "UA"){
    countryCode<- "USSR"
  }
})
countryMatch <-  match.map("world", country)
#color code any states with visit data as light blue
colorCountry <- sapply(countryMatch, function(c){
 if(!is.na(c)) c <- "#C6DBEF"
 else c <- "#FFFFFF"
})
m <- map('world',plot=FALSE)
map('world',proj='azequalarea',orient=c(41,-74,0), boundary=TRUE,
col=colorCountry,fill=TRUE)
map.grid(m,col="blue", label=FALSE, lty=2, pretty=FALSE)
map.scale()
```

```{r fig.width=10, fig.height=7}
#isolate the US data, scrub any unknown states
usData <- logData[logData$state != "XX", ]
usData$state <- apply(as.matrix(usData$state), 1, function(s){
  #trim the abbreviation of whitespaces
  s <- gsub("(^ +)|( +$)", "", s)
```

```
  s <- state.name[grep(s, state.abb)]
})
s <- map('state',plot=FALSE)
states <- unique(usData$state)
stateMatch <- match.map("state", states)
#color code any states with visit data as light blue
colorMatch <- sapply(stateMatch, function(s){
 if(!is.na(s)) s <- "#C6DBEF"
 else s <- "#FFFFFF"
})
map("state", resolution = 0,lty = 0,projection = "azequalarea",
col=colorMatch,fill=TRUE)
map("state", col = "black",fill=FALSE,add=TRUE,lty=1,lwd=1,projection="azeq
ualarea")
map.scale()
```

This code produces the output shown in Figure 5-12. I have also made this R script available in the code download for this book.

Visualizing Spatial Data from Access Logs

```
#chart worldwide visit data
#unfortunately there is no state.name equivalent for countries so we must check
#the explicit country names. In the us states below we are able to accomplish this much
#more efficiently
country <- unique(logData$country)
country <- sapply(country, function(countryCode){
  #trim whitespaces from the country code
  countryCode <- gsub("(^ +)|( +$)", "", countryCode)
  if(countryCode == "US"){
    countryCode<- "USA"
  }else if(countryCode == "AU"){
    countryCode<- "Australia"
  }}
)
countryMatch <-  match.map("world2", country)
colorCountry <- sapply(countryMatch, function(c){
 if(!is.na(c)) c <- "#C6DBEF"
 else c <- "#FFFFFF"
})
map('world', proj='azequalarea', orient=c(41,-74,0), boundary=TRUE, col=colorCount
ry, fill=TRUE)
m <- map('world',plot=FALSE)
map.grid(m, col="blue", label=FALSE, lty=2, pretty=TRUE)
map.scale()
```

Figure 5-12. *Data maps in R Markdown*

149

```
#isolate the US data, scrub any unknown states
usData <- logData[logData$state != "XX", ]
usData$state <- apply(as.matrix(usData$state), 1, function(s){
  #trim the abbreviation of whitespaces
  s <- gsub("(^ +)|( +$)", "", s)
  s <- state.name[grep(s, state.abb)]
})
s <- map('state',plot=FALSE)
states <- unique(usData$state)
stateMatch <- match.map("state", states)
#color code any states with visit data as light blue
colorMatch <- sapply(stateMatch, function(s){
 if(!is.na(s)) s <- "#C6DBEF"
 else s <- "#FFFFFF"
})
map("state", resolution = 0,lty = 0,projection = "azequalarea", col=colorMatch,fil
l=TRUE)
map("state", col = "black",fill=FALSE,add=TRUE,lty=1,lwd=1,projection="azequalarea
")
map.scale()
```

Figure 5-12. (*continued*)

Summary

This chapter discussed parsing access logs to produce data map visualizations. You looked at both global country data in your maps and more localized state data. This is the first taste of how you can begin to bring usage data to life.

The next chapter looks at bug backlog data in the context of time series charts.

CHAPTER 6

Visualizing Data over Time

The last chapter discussed using access logs to create data maps representing the geographic location of users. We used the `map` and `mapproj` (for map projections) packages to create these visualizations.

This chapter explores creating time series charts, which are graphs that compare changes in values over time. They are generally read left to right with the x-axis representing some measure of time and the y-axis representing the range of values. This chapter discusses visualizing defects over time.

Tracking defects over time allows us to identify not only spikes in issues but also larger patterns in workflows, especially when we include more granular details such as bug criticality and include cross-referencing data such as dates for events like start and end of iteration. We begin to expose trends such as when during an iteration bugs get opened, when most of the blocker bugs get opened, or what iterations produce the highest number of bugs. This kind of self-evaluation and reflection are what allow us to identify and focus attention on blind spots or areas of improvement. It also allows us to recognize victories in a larger scope that might be missed when viewing the daily numbers without context.

A case in point: recently our organization set a larger group goal of achieving a certain bug number by the end of the year, a percent of the total open bugs that we had open at the beginning of the year. With our peers and our management staff, we coached all the developers, created process improvements, and won hearts and minds for this goal. At the end of the year, the number of bugs we had remaining open was about the same as when we had started. We were confused and concerned. But when we summed the daily numbers, we realized that we had achieved something larger than we anticipated: we actually opened one-third fewer bugs overall year over year from the previous year. This was huge and would easily have been missed if we weren't looking at the data with a critical eye to the larger picture.

© Tom Barker, Jon Westfall 2022
T. Barker and J. Westfall, *Pro Data Visualization Using R and JavaScript,*
https://doi.org/10.1007/978-1-4842-7202-2_6

Gathering Data

The first step of creating a defect time series chart is to decide on a time period that we want to look at and gather the data. This means getting an export of all the bugs for a given time period.

This step is completely dependent on the bug tracing software that you may use. Maybe you use HP's Quality Center because it makes sense with the rest of your organization's testing needs (such as being able to work with LoadRunner). Maybe you use a hosted web-based solution such as Rally because you get defect management bundled in with your user story and release tracking. Maybe you have your own installation of Bugzilla because it's open and free.

Whatever the case, all defect management software has a way to export your current bug list. Depending on the defect-tracking software used, you can export to a flat file, such as a comma or tab-separated file. The software can also allow access to its contents via an API so you can create a script that accesses the API and exposes the content.

Either way, there are two important main cases when looking at bugs over time:

- Running total of bugs by date

- New bugs by date

For either of these cases, the minimum fields that we care about when we export from the bug-tracking software are the following:

- Date opened

- Defect id

- Defect status

- Severity of defect

- Description of the defect

The exported bug data should look something like this:

```
Date, ID, Severity, Status, Summary
6/7/20,DE45091,Minor,Open,videos not playing
8/21/20,DE45092,Blocker,Open,alignment off
3/7/20,DE45093,Moderate,Closed,monsters attacking
```

Let's process the data to be able to visualize it.

Data Analysis with R

The first thing is to read in and order the data. Assuming that data is exported to a flat file named allbugs.csv, we can read in the data as follows (we have provided sample data for it at http://jonwestfall.com/data/allbugs.csv):

```
bugExport <- "/Applications/MAMP/htdocs/allbugs.csv"
bugs <- read.table(bugExport, header=TRUE, sep=",")
```

Let's order the data frame by date. To do this, we have to convert the Date column, which is read in as a string, into a Date object using the as.Date() function. The as. Date() function accepts several symbols to signify how to read and structure the date object, as shown in Table 6-1.

Table 6-1. *as.Date( ) function symbols*

Symbol	Meaning
%m	Numeric month
%b	Month name as string, abbreviated
%B	Full month name as string
%d	Numeric day
%a	Weekday as abbreviated string
%A	Full weekday as string
%y	Year as two-digit number
%Y	Year as four-digit number

So for the date "04/01/2013", we pass in "%m/%d/%Y"; for "April 01, 13", we pass in "%B %d, %Y". You can see how the pattern matches up:

```
as.Date(bugs$Date,"%m/%d/%y")
```

We'll use the converted date in the order() function, which returns a list of index numbers from the bugs data frame, corresponding with the correct way to order the values in the data frame:

```
> order(as.Date(bugs$Date,"%m/%d/%y"))
  [1] 127  90 187 112  13 119 137 101  37  53  52  67 125   4  81  93 136
  3  55  62  33  25 130  75  85  28
 [27]  44 159 126 107  30 191  80 124  36 104  18  24  82  20  21  34  56
147  29 156  16  59  51 139   1 123
 [53] 113 146 148   5 103  43  83  23 173  11 168  99  35   7 192  42 142
121   9  69   2 171  60  94 164  17
 [79]  91  84 178  96 105   8 110  39 177 109  97 120 135  58  79  15
111  49 117  50  57  92 129 114 145 158
[105] 116 151 143 162  31  73  77 182  26  74 195  10  48  88  76 183 115
184 189 108  61 174 144 186  12 134
[131] 157  41  86  27 175   6 165  46 118 188  65 141  22 169 190  72  66
154  40  47  64 166  14  87  95 155
[157] 193 133 179  54 140 128  89 102 161  63  45  78 138 180 149 185
106  38 181 172 176 153 160 150 170 122
[183] 194 100 167  68  98 132  70 152  19 163  71  32 131
```

Finally, we'll use the results of the order() function as the indexes of the bugs data frame and pass the results back into the bugs data frame:

```
bugs <- bugs[order(as.Date(bugs$Date," %m/%d/%y ")),]
```

This code reorders the bugs data frame based on the order of the indexes returned in the order() function. It will be handy when we begin to slice up the data. The data frame should now be a chronologically ordered list of bugs, which looks like the following:

```
> bugs
        Date       ID Severity Status                 Summary
127  1/3/20 DE45217    Minor   Open      Mug of coffee empty
90   1/4/20 DE45180    Minor Closed mug of coffee destroyed
187  1/5/20 DE45277    Minor   Open             Zerg attack
112  1/9/20 DE45202  Blocker Closed                 Monkeys
13  1/12/20 DE45103    Minor   Open      Mug of coffee empty
119 1/13/20 DE45209  Blocker Closed      The plague occurred
```

```
write.table(bugs, col.names=TRUE, row.names=FALSE, file="allbugsOrdered.
csv", quote = FALSE, sep = ",")
```

This will come in handy later when we look at this data in D3.

Calculating the Bug Count

Next, we will calculate the total bug count by date. This will show how many new bugs are opened by day.

To do this, we pass bugs$Date into the table() function, which builds a data structure of counts of each date in the bugs data frame:

```
totalBugsByDate <- table(bugs$Date)
```

So the structure of totalBugsByDate looks like the following:

```
> totalBugsByDate
```

1/11/21	1/12/20	1/12/21	1/13/20	1/17/21	1/18/21	1/2/21	1/21/20	1/22/20
1	1	3	1	2	1	1	1	1
1/24/20	1/24/21	1/25/20	1/27/21	1/29/21	1/3/20	1/4/20	1/5/20	1/5/21
1	1	1	1	1	1	1	1	1
1/9/20	10/1/20	10/10/20	10/15/20	10/16/20	10/18/20	10/21/20	10/25/20	10/26/20
1	1	1	1	1	2	2	1	1
10/29/20	10/30/20	10/6/20	11/17/20	11/18/20	11/19/20	11/21/20	11/23/20	11/26/20
2	1	1	1	1	1	1	1	2
11/4/20	11/8/20	12/14/20	12/15/20	12/17/20	12/21/20	12/22/20	12/23/20	12/24/20
2	1	2	1	1	1	2	1	1
12/27/20	12/29/20	12/3/20	12/31/20	2/12/21	2/13/21	2/14/20	2/15/20	2/15/21
1	1	1	1	1	1	1	1	1
2/16/20	2/22/21	2/24/20	2/25/21	2/26/21	2/28/21	2/3/21	2/4/21	2/8/21
1	2	1	1	2	1	1	1	1
3/1/20	3/1/21	3/11/21	3/14/21	3/17/21	3/2/20	3/2/21	3/22/20	3/23/21
2	1	3	1	1	1	1	2	1
3/24/20	3/25/21	3/26/20	3/28/20	3/3/21	3/31/20	3/31/21	3/6/21	3/7/20

157

1	1	1	1	1	1	1	1	1
3/7/21	4/12/21	4/13/20	4/15/21	4/18/21	4/19/21	4/20/20	4/25/20	4/26/21

1	1	1	1	2	1	1	1	1
4/27/20	4/29/21	4/4/20	4/5/21	4/7/20	4/8/20	5/1/20	5/10/20	5/11/21

1	1	1	3	1	2	2	1	1
5/12/20	5/14/21	5/16/21	5/17/20	5/17/21	5/2/21	5/20/20	5/20/21	5/22/20

2	1	1	1	1	1	1	2	2
5/24/21	5/25/20	5/26/21	5/27/20	5/27/21	5/28/20	5/28/21	5/29/21	5/30/20

1	1	1	1	1	1	1	2	1
5/31/20	5/6/20	5/8/20	6/11/20	6/11/21	6/14/20	6/16/21	6/2/21	6/20/20

1	1	1	1	1	1	2	1	1
6/28/20	6/3/20	6/3/21	6/4/20	6/4/21	6/6/21	6/7/20	6/7/21	6/8/21

1	1	1	1	1	1	2	1	1
6/9/21	7/14/20	7/18/20	7/2/20	7/22/20	7/23/20	7/25/20	7/28/20	7/29/20

1	1	2	1	1	1	1	1	1
7/9/20	8/10/20	8/17/20	8/2/20	8/21/20	8/22/20	8/23/20	8/24/20	8/26/20

1	1	2	1	1	1	1	2	1
8/27/20	8/28/20	8/29/20	8/3/20	8/6/20	9/10/20	9/11/20	9/14/20	9/16/20

1	1	1
9/2/20	9/21/20	9/8/20

Let's plot this data out to get an idea of how many bugs are opened each day:

```
plot(totalBugsByDate, type="l", main="New Bugs by Date", col="red",
ylab="Bugs")
```

This code creates the chart shown in Figure 6-1.

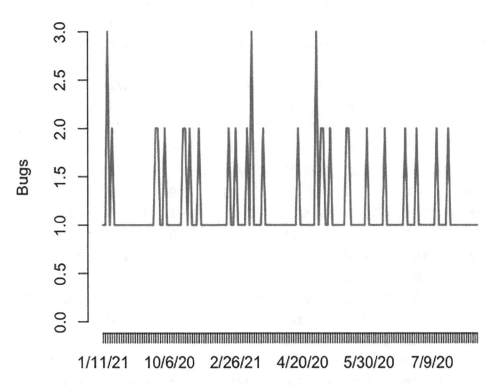

Figure 6-1. *Time series of new bugs by date*

Now that we have a count of how many bugs are generated each day, we can get a cumulative sum by using the cumsum() function. It takes the new bugs opened each day and creates a running sum of them, updating the total each day. It allows us to generate a trend line for the cumulative count of bugs over time.

```
> runningTotalBugs <- cumsum(totalBugsByDate)
>
> runningTotalBugs
 1/11/21  1/12/20  1/12/21  1/13/20  1/17/21  1/18/21   1/2/21  1/21/20  1/22/20
       1        2        5        6        8        9       10       11       12
 1/24/20  1/24/21  1/25/20  1/27/21  1/29/21   1/3/20   1/4/20   1/5/20   1/5/21
      13       14       15       16       17       18       19       20       21
  1/9/20  10/1/20 10/10/20 10/15/20 10/16/20 10/18/20 10/21/20 10/25/20 10/26/20
      22       23       24       25       26       28       30       31       32
10/29/20 10/30/20  10/6/20 11/17/20 11/18/20 11/19/20 11/21/20 11/23/20 11/26/20
```

34	35	36	37	38	39	40	41	43
11/4/20	11/8/20	12/14/20	12/15/20	12/17/20	12/21/20	12/22/20	12/23/20	12/24/20

45	46	48	49	50	51	53	54	55
12/27/20	12/29/20	12/3/20	12/31/20	2/12/21	2/13/21	2/14/20	2/15/20	2/15/21

56	57	58	59	60	61	62	63	64
2/16/20	2/22/21	2/24/20	2/25/21	2/26/21	2/28/21	2/3/21	2/4/21	2/8/21

65	67	68	69	71	72	73	74	75
3/1/20	3/1/21	3/11/21	3/14/21	3/17/21	3/2/20	3/2/21	3/22/20	3/23/21

77	78	81	82	83	84	85	87	88
3/24/20	3/25/21	3/26/20	3/28/20	3/3/21	3/31/20	3/31/21	3/6/21	3/7/20

89	90	91	92	93	94	95	96	97
3/7/21	4/12/21	4/13/20	4/15/21	4/18/21	4/19/21	4/20/20	4/25/20	4/26/21

98	99	100	101	103	104	105	106	107
4/27/20	4/29/21	4/4/20	4/5/21	4/7/20	4/8/20	5/1/20	5/10/20	5/11/21

108	109	110	113	114	116	118	119	120
5/12/20	5/14/21	5/16/21	5/17/20	5/17/21	5/2/21	5/20/20	5/20/21	5/22/20

122	123	124	125	126	127	128	130	132
5/24/21	5/25/20	5/26/21	5/27/20	5/27/21	5/28/20	5/28/21	5/29/21	5/30/20

133	134	135	136	137	138	139	141	142
5/31/20	5/6/20	5/8/20	6/11/20	6/11/21	6/14/20	6/16/21	6/2/21	6/20/20

143	144	145	146	147	148	150	151	152
6/28/20	6/3/20	6/3/21	6/4/20	6/4/21	6/6/21	6/7/20	6/7/21	6/8/21

153	154	155	156	157	158	160	161	162
6/9/21	7/14/20	7/18/20	7/2/20	7/22/20	7/23/20	7/25/20	7/28/20	7/29/20

163	164	166	167	168	169	170	171	172
7/9/20	8/10/20	8/17/20	8/2/20	8/21/20	8/22/20	8/23/20	8/24/20	8/26/20

173	174	176	177	178	179	180	182	183
8/27/20	8/28/20	8/29/20	8/3/20	8/6/20	9/10/20	9/11/20	9/14/20	9/16/20

184	185	186	187	188	189	190	191	192
9/2/20	9/21/20	9/8/20						

193	194	195

This is exactly what we need to now plot out the way the bug backlog grows or shrinks each day. To do that, let's pass runningTotalBugs to the plot() function. We set the type to "l" to signify that we are creating a line chart and then name the chart Cumulative Defects Over Time. In the plot() function, we also turn the axes off so that we can draw custom axes for this chart. We will want to draw custom axes so that we can specify the dates as the x-axis labels.

To draw custom axes, we use the axis() function. The first parameter in the axis() function is a number that tells R where to draw the axis.

- 1 corresponds to the x-axis at the bottom of the chart.

- 2 to the left of the chart.

- 3 to the top of the chart.

- 4 to the right of the chart.

```
plot(runningTotalBugs, type="l", xlab="", ylab="", pch=15, lty=1,
col="red", main="Cumulative Defects Over Time", axes=FALSE)
axis(1, at=1: length(runningTotalBugs), lab= row.names(totalBugsByDate))
axis(2, las=1, at=10*0:max(runningTotalBugs))
```

Note that the plot type is set to a lowercase L, not an uppercase i or 1. This code creates the time series chart shown in Figure 6-2.

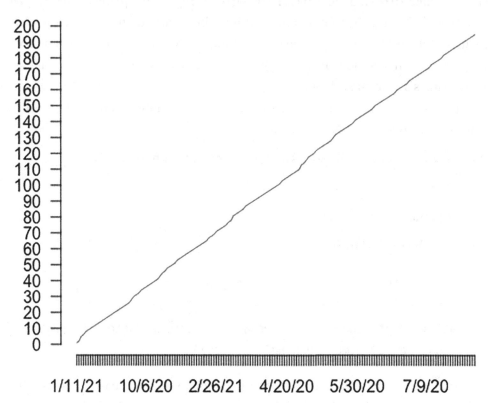

Figure 6-2. *Cumulative defects over time*

This shows the progressively increasing bug backlog, by date.

The complete R code so far is as follows:

```
bugExport <- "allbugs.csv"
bugs <- read.table(bugExport, header=TRUE, sep=",")
as.Date(bugs$Date,"%m/%d/%y")
order(as.Date(bugs$Date,"%m/%d/%y"))
bugs <- bugs[order(as.Date(bugs$Date," %m/%d/%y ")),]
write.table(bugs, col.names=TRUE, row.names=FALSE, file="allbugsOrdered.
csv", quote = FALSE, sep = ",")

totalBugsByDate <- table(bugs$Date)
plot(totalBugsByDate, type="l", main="New Bugs by Date", col="red",
ylab="Bugs")
```

```
runningTotalBugs <- cumsum(totalBugsByDate)
runningTotalBugs
plot(runningTotalBugs, type="l", xlab="", ylab="", pch=15, lty=1,
col="red", main="Cumulative Defects Over Time", axes=FALSE)
axis(1, at=1: length(runningTotalBugs), lab= row.names(totalBugsByDate))
axis(2, las=1, at=10*0:max(runningTotalBugs))
```

Let's take a look at the criticality of the bugs, which shows not just when the bugs are opened but also when the most severe (or non-severe) bugs are being opened.

Examining the Severity of the Bugs

Remember that when we exported the bug data, we included the Severity field, which indicates the level of criticality of each bug. Each team and organization might have their own classification of severity, but generally they include these:

- **Blockers** are bugs so severe that they prevent the launch of a body of work. They generally have broken functionality or are missing sections of a widely used feature. They can also be discrepancies with contractually or legally binding features such as closed captioning or digital rights protection.

- **Moderates** are bugs that are severe but not so damaging that they gate a release. They can have broken functionality of less-used features. The scope of accessibility, or how widely used a feature is, is usually a determining factor between making a bug a blocker or a critical.

- **Minors** are bugs with very minimal if any impact and might not even be noticeable to an end user.

To break out the bugs by severity, we simply call the table() function, just as we did to break out bugs out by date, but this time add in the Severity column as well:

```
bugsBySeverity <- table(factor(bugs$Date),bugs$Severity)
```

163

This code creates a data structure that looks like so:

	Blocker	Minor	Moderate
1/11/21	0	1	0
1/12/20	0	1	0
1/12/21	1	2	0
1/13/20	1	0	0
1/17/21	2	0	0
1/18/21	0	0	1
1/2/21	0	1	0
1/21/20	1	0	0
1/22/20	1	0	0
1/24/20	0	1	0

We can then plot this data object. The way we do this is to use the plot() function to create a chart for one of the columns and then use the lines() function to draw lines on the chart for the remaining columns:

```
plot(bugsBySeverity[,3], type="l", xlab="", ylab="", pch=15, lty=1,
col="orange", main="New Bugs by Severity and Date", axes=FALSE)
lines(bugsBySeverity[,1], type="l", col="red", lty=1)
lines(bugsBySeverity[,2], type="l", col="yellow", lty=1)
axis(1, at=1: length(runningTotalBugs), lab= row.names(totalBugsByDate))
axis(2, las=1, at=0:max(bugsBySeverity[,3]))
legend("topleft", inset=.01, title="Legend", colnames(bugsBySeverity),
lty=c(1,1,1), col= c("red", "yellow", "orange"))
```

This code produces the chart shown in Figure 6-3.

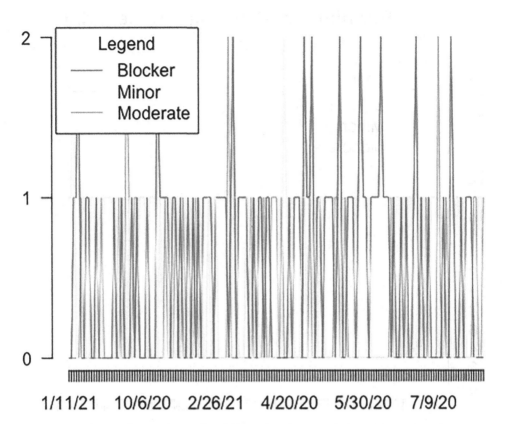

Figure 6-3. *Our plot() and lines() functions drawing the chart of bugs by severity*

This is great, but what if we want to see the cumulative bugs by severity? We can simply use the preceding R code, but instead of plotting out the columns, we can plot out the cumulative sum of each column:

```
plot(cumsum(bugsBySeverity[,3]), type="l", xlab="", ylab="", pch=15, lty=1,
col="orange", main="Running Total of Bugs by Severity", axes=FALSE)
lines(cumsum(bugsBySeverity[,1]), type="l", col="red", lty=1)
lines(cumsum(bugsBySeverity[,2]), type="l", col="yellow", lty=1)
axis(1, at=1: length(runningTotalBugs), lab= row.names(totalBugsByDate))
axis(2, las=1, at=0:max(cumsum(bugsBySeverity[,3])))
legend("topleft", inset=.01, title="Legend", colnames(bugsBySeverity),
lty=c(1,1,1), col= c("red", "yellow", "orange"))
```

This code produces the chart shown in Figure 6-4.

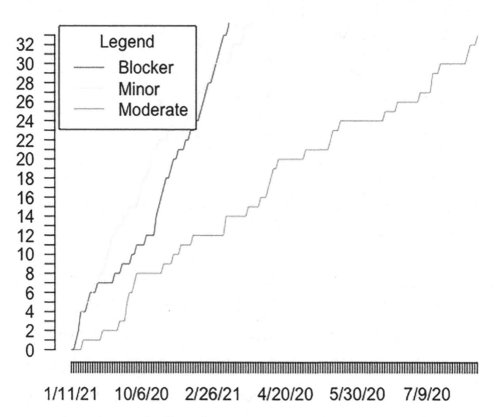

Figure 6-4. *Running total of bugs by severity*

Adding Interactivity with D3

The previous example is a great way to visualize and disseminate information around the creation of defects. But what if we could take it a step further and allow the consumers of our visualizations to dive deeper into the data points that interest them? Say we wanted to allow the user to mouse over a particular point in a time series and see a list of all the bugs that make up that data point. We can do just that with D3; let's walk through it and find out how.

First, let's create a new file with the base HTML skeletal structure with a reference to D3.js and save it as timeseriesGranular.htm. We'll want to use the older version of D3 for this example—version 3 (d3.v3.js, available in the code download for this book), in that it allowed for a bit more flexibility and step-by-step building than the newer code structure.

```
<html>
<head></head>
<body>
<script src="d3.v3.js"></script>
</body>
</html>
```

Next, we set some preliminary data in a new `script` tag. We create an object to hold margin data for the graphic, as well as height and width. We also create a D3 time formatter to convert the dates that are read in from string to a native `Date` object.

```
<script>
var margin = {top: 20, right: 20, bottom: 30, left: 50},
 width = 960 - margin.left - margin.right,
 height = 500 - margin.top - margin.bottom;
var parseDate = d3.timeFormat("%m/%d/%y").parse;
</script>
```

Reading in the Data

We add in some code to read in the data (the `allbugsOrdered.csv` file that was output from R earlier). Recall that this file contains the entire bug data ordered by date.

We use the `d3.csv()` function to read this file:

- The first parameter is the path to the file.

- The second parameter is the function to execute once the data is read in. It is in this anonymous function that we add most of the functionality, or at least the functionality that is dependent on having data to process.

The anonymous function accepts two parameters:

- The first catches any errors that may occur.

- The second is the contents of the file being read in.

In the function, we first loop through the contents of the data and use the date formatter to convert all the values in the Date column to a native JavaScript Date object:

```
d3.csv("allbugsOrdered.csv", function(error, data) {
        data.forEach(function(d) {
        d.Date = parseDate(d.Date);
});
});
```

If we were to console.log() the data, it would be an array of objects that look like Figure 6-5.

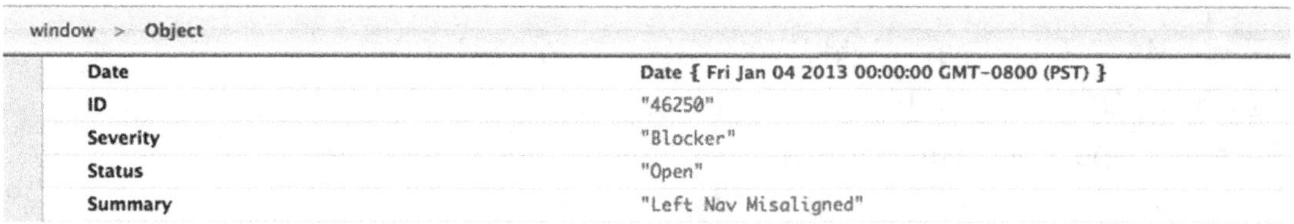

window > Object	
Date	Date { Fri Jan 04 2013 00:00:00 GMT-0800 (PST) }
ID	"46250"
Severity	"Blocker"
Status	"Open"
Summary	"Left Nav Misaligned"

Figure 6-5. *Our bug data object*

Within the anonymous function but after the loop, we use the d3.nest() function to create a variable that holds the bug data grouped by date. We name this variable nested_data:

```
nested_data = d3.nest()
.key(function(d) { return d.Date; })
.entries(data);
```

The nested_data variable is now a tree structure—specifically a list that is indexed by date, and each index has a list of bugs. If we were to console.log() nested_data, it would be an array of objects that look like Figure 6-6.

key	"Fri Jan 04 2013 00:00:00 GMT-0800 (PST)"
▼ values	[Object { Date=Date, ID="46250", Severity="Critical", more... }, Object { Date=Date, ID="46253", Severity=" Minor", more... }]
▼ 0	Object { Date=Date, ID="46250", Severity="Critical", more... }
Date	Date { Fri Jan 04 2013 00:00:00 GMT-0800 (PST) }
ID	"46250"
Severity	"Critical"
Status	"Open"
Summary	"Incorrect icon"
▼ 1	Object { Date=Date, ID="46253", Severity=" Minor", more... }
Date	Date { Fri Jan 04 2013 00:00:00 GMT-0800 (PST) }
ID	"46253"
Severity	" Minor"
Status	" Deferred"
Summary	"Homepage not loading"

Figure 6-6. *The array containing our bug data objects*

Drawing on the Page

We are ready to start drawing to the page. So let's step out of the callback function and go to the root of the `script` tag and write out the SVG tag to the page by using the margins, width, and height that were defined previously:

```
var svg = d3.select("body").append("svg")
 .attr("width", width + margin.left + margin.right)
 .attr("height", height + margin.top + margin.bottom)
        .append("g")
 .attr("transform", "translate(" + margin.left + "," + margin.top + ")");
```

This is the container in which we draw the axes and the trend lines.

Still at the root level, we add a D3 `scale` object for both the x- and y-axes, using the `width` variable for the x-axis range and the `height` variable for the y-axis range. We add the x- and y-axes at the root level, passing in their respective scale objects and orienting them at the bottom and left.

```
var xScale = d3.time.scale()
 .range([0, width]);
var yScale= d3.scale.linear()
 .range([height, 0]);
var xAxis = d3.svg.axis()
 .scale(xScale)
 .orient("bottom");
var yAxis = d3.svg.axis()
 .scale(yScale)
 .orient("left");
```

But they still aren't showing on the page. We need to return to the anonymous function that we created in the `d3.csv()` call and add the `nested_data` list that we created as the domain data for the newly created scales:

```
xScale.domain(d3.extent(nested_data, function(d) { return new Date(d.key); }));
yScale.domain(d3.extent(nested_data, function(d) { return d.values.length; }));
```

169

From here, we need to generate the axes. We do this by adding and selecting an SVG g element, used for generic grouping, and adding this selection to the xAxis() and yAxis() D3 functions. This also goes in the anonymous callback function that gets invoked when the data is loaded.

We also need to transform the x-axis by adding the height of the chart so that it is drawn at the bottom of the graph:

```
svg.append("g")
 .attr("transform", "translate(0," + height + ")")
 .call(xAxis);
svg.append("g")
 .call(yAxis)
```

This creates the start of the chart with meaningful axes shown in Figure 6-7.

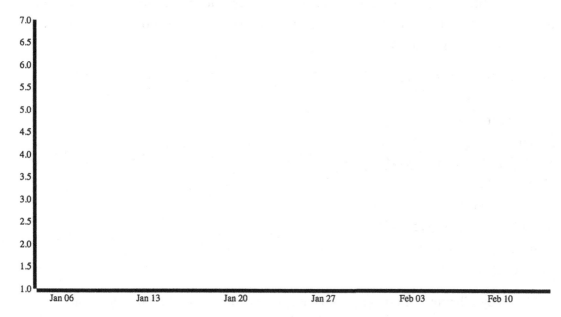

Figure 6-7. *Time series beginning to form; x- and y-axes but no line yet*

The trend line needs to be added. Back at the root level, let's create a variable named line to be an SVG line. Assume for a minute that we have already set the data property for the line. We haven't yet, but we will in a minute. For the x value of the line, we will have a function that returns the date filtered through the xScale scale object. For the y value of the line, we will create a function that returns the bug count values run through the yScale scale object.

```
var line = d3.svg.line()
 .x(function(d) { return xScale(new Date(d.key)); })
 .y(function(d) { return yScale(d.values.length); });
```

Next, we return to the anonymous function that processes the data. Right below the added axes, we will append an SVG path. We set the nested_data variable as the datum for the path and the newly created line object as the d attribute. For reference, the d attribute is where we specify path descriptions. See here for documentation around the d attribute: https://developer.mozilla.org/en-US/docs/SVG/Attribute/d.

```
svg.append("path")
 .datum(nested_data)
 .attr("d", line);
```

We can now start to see something in a browser. The code so far should look like so:

```
<!DOCTYPE html>
<head>
<meta charset="utf-8">
</head>
<body>
        <script src="d3.v3.js"></script>
<script>
var margin = {top: 20, right: 20, bottom: 30, left: 50},
 width = 960 - margin.left - margin.right,
 height = 500 - margin.top - margin.bottom;
var parseDate = d3.time.format("%m-%d-%Y").parse;
var xScale = d3.time.scale()
 .range([0, width]);
var yScale = d3.scale.linear()
 .range([height, 0]);
var xAxis = d3.svg.axis()
        .scale(xScale)
        .orient("bottom");
var yAxis = d3.svg.axis()
        .scale(yScale)
        .orient("left");
```

```
var line = d3.svg.line()
 .x(function(d) { return xScale(new Date(d.key)); })
        .y(function(d) { return yScale(d.values.length); });
var svg = d3.select("body").append("svg")
 .attr("width", width + margin.left + margin.right)
 .attr("height", height + margin.top + margin.bottom)
       . .append("g")
 .attr("transform", "translate(" + margin.left + "," + margin.top + ")");
d3.csv("allbugsOrdered.csv", function(error, data) {
        data.forEach(function(d) {
                d.Date = parseDate(d.Date);
        });
 nested_data = d3.nest()
                .key(function(d) { return d.Date; })
                .entries(data);
        xScale.domain(d3.extent(nested_data, function(d) { return new
        Date(d.key); }));
        yScale.domain(d3.extent(nested_data, function(d) { return
        d.values.length; }));
        svg.append("g")
        .attr("transform", "translate(0," + height + ")")
        .call(xAxis);
                svg.append("g")
                 .call(yAxis);
                svg.append("path")
                 .datum(nested_data)
                 .attr("d", line);
});
</script>
</body>
</html>
```

This code produces the graphic shown in Figure 6-8.

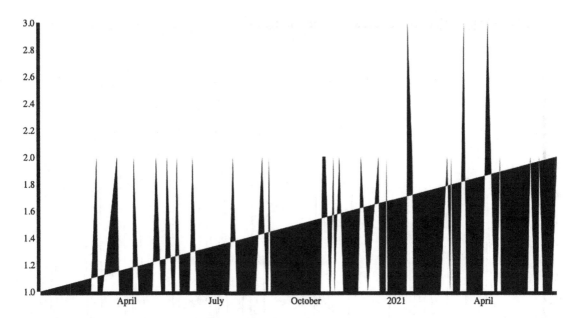

Figure 6-8. *Time series with line data but incorrect fill*

But this isn't quite right. The shading of the path is based on the browser's best guess of intent, shading what it perceives to be the closed areas. Let's use CSS to explicitly turn off shading and instead set the color and width of the path line:

```
<style>
.trendLine {
 fill: none;
 stroke: #CC0000;
 stroke-width: 1.5px;
}
</style>
```

We created a style rule for any element on the page with the class trendLine. Let's next add the class to the SVG path in the same block of code in which we create the path:

```
Svg.append("path")
 .datum(nested_data)
 .attr("d", line)
 .attr("class", "trendLine");
```

This code produces the chart shown in Figure 6-9.

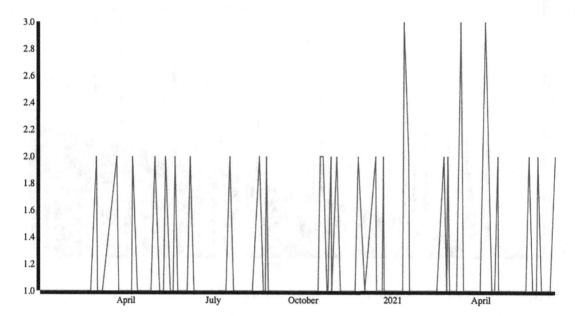

Figure 6-9. *Time series with corrected line but unstyled axes*

Looking much better! There are some minor things we should change, such as adding text labels to the y-axis and trimming the width of the axis lines to make them neater:

```
.axis path{
 fill: none;
 stroke: #000;
 shape-rendering: crispEdges;
}
```

This will give us tighter-looking axes. We just need to apply the style to the axes when we create them:

```
svg.append("g")
 .attr("transform", "translate(0," + height + ")")
 .call(xAxis)
 .attr("class", "axis");
svg.append("g")
 .call(yAxis)
 .attr("class", "axis");
```

The results can be seen in Figure 6-10.

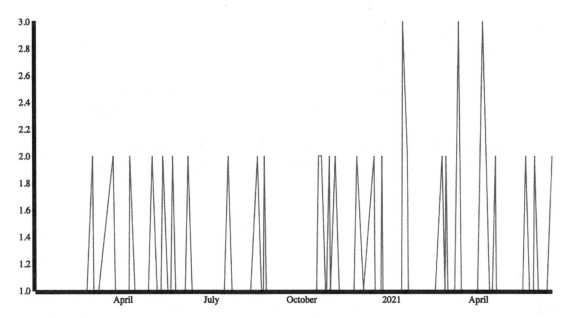

Figure 6-10. *Time series updated with styled axes*

This is great so far, but it shows no real benefit from doing this in R. In fact, we wrote quite a bit of additional code just to get parity and didn't even do any data cleaning that we did in R.

The real benefit of using D3 is adding interactivity.

Adding Interactivity

Say we have this time series of new bugs, and we were curious what the bugs were in that large spike in mid-February. By taking advantage of the fact that we are working in HTML and JavaScript, we can extend this functionality by adding in a tooltip box that lists the bugs for each date.

To do this, we first should create obvious areas in which users can mouse over, such as red circles at each data point or discrete date. To do that, we simply need to create SVG circles right below where we added in the path, in the anonymous function that is fired when the external data is read in. We set the `nested_data` variable as the `data` attribute of the circles, make them red with a radius of 3.5, and set their x and y attributes to be tied to the date and bug totals, respectively:

```
svg.selectAll("circle")
.data(nested_data)
.enter().append("circle")
        .attr("r", 3.5)
        .attr("fill", "red")
        .attr("cx", function(d) { return xScale(new Date(d.key)); })
        .attr("cy", function(d) { return yScale(d.values.length);})
```

This code updates the existing time series so it looks like Figure 6-11. These red circles are now areas of focus in which users can mouse over and see additional information.

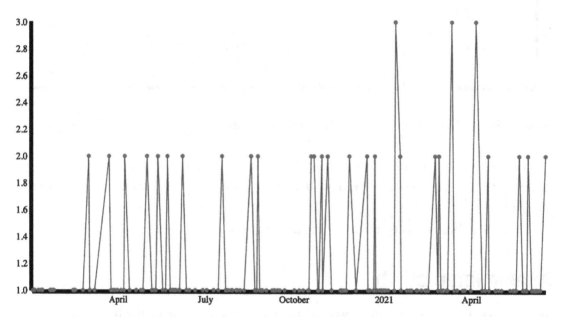

Figure 6-11. *Circles added to each data point on the line*

Let's next code up a div to act as the tooltip that we will show with relevant bug data. To do this, we will create a new div, right below where we created the line variable at the root of the script tag. We do this in D3 once again by selecting the body tag and appending a div to it, giving it a class and id of tooltip—both so that we can have the tooltip style apply to it (which we will create in just a minute) and so we can interact with it by ID later on in the chapter. We will have it hidden by default. We will store a reference to this div in a variable that we will call tooltip.

```
var tooltip = d3.select("body")
 .append("div")
 .attr("class", "tooltip")
 .attr("id", "tooltip")
 .style("position", "absolute")
 .style("z-index", "10")
 .style("visibility", "hidden");
```

We next need to style this div using CSS. We adjust the opacity to be only 75 percent visible, so that when the tooltip shows up over a trend line, we can see the trend line behind it. We align the text, set the font size, make the div have a white background, and give it rounded corners.

```
.tooltip{
        opacity: .75;
        text-align:center;
        font-size:12px;
        width:100px;
        padding:5px;
        border:1px solid #a8b6ba;
        background-color:#fff;
        margin-bottom:5px;
        border-radius: 19px;
        -moz-border-radius: 19px;
        -webkit-border-radius: 19px;
}
```

We next have to add a mouseover event handler to the circles to populate the tooltip with information and unhide the tooltip. To do this, we return to the block of code in which we created the circles and add in a mousemove event handler that fires off an anonymous function.

Inside the anonymous function, we overwrite the innerHTML of the tooltip to display the date of the current red circle and how many bugs are associated with that date. We then loop through that list of bugs and write out the ID of each bug.

```
svg.selectAll("circle")
 .data(nested_data)
 .enter().append("circle")
```

```
.attr("r", 3.5)
.attr("fill", "red")
.attr("cx", function(d) { return xScale(new Date(d.key)); })
.attr("cy", function(d) { return yScale(d.values.length);})
.on("mouseover", function(d){
document.getElementById("tooltip").innerHTML = d.key + " " + d.values.
length + " bugs<br/>";
 for(x=0;x<d.values.length;x++){
document.getElementById("tooltip").innerHTML += d.values[x].ID + "<br/>";
 }
tooltip.style("visibility", "visible");
 })
```

If we want to take this even further, we can create links for each bug ID that link back to the bug-tracking software, list descriptions of each bug, and if the bug-tracking software has an API to interface with, we can even have form fields that could let us update bug information right from this tooltip. Only our imagination and the tools available to us limit the possibilities of how far we can extend this concept.

Finally, we add a mousemove event handler to the red circles so that we can reposition the tooltip contextually whenever the users mouse over a red circle. To do this, we use the d3.mouse object to get the current mouse coordinates. We use these coordinates to simply reposition the tooltip with CSS. So we don't cover the red circle with the tooltip, we offset the top property by 25 pixels and the left property by 75 pixels.

```
svg.selectAll("circle")
 .data(nested_data)
 .enter().append("circle")
 .attr("r", 3.5)
 .attr("fill", "red")
 .attr("cx", function(d) { return xScale(new Date(d.key)); })
 .attr("cy", function(d) { return yScale(d.values.length);})
 .on("mouseover", function(d){
document.getElementById("tooltip").innerHTML = d.key + " " + d.values.
length + " bugs<br/>";
for(x=0;x<d.values.length;x++){
document.getElementById("tooltip").innerHTML += d.values[x].ID + "<br/>";
```

```
}
tooltip.style("visibility", "visible");
})
.on("mousemove", function(){
return tooltip.style("top", (d3.mouse(this)[1] + 25)+"px").style("left",
(d3.mouse(this)[0] + 70)+"px");
});
```

A tooltip should display when the mouse hovers over one of the red circles (see Figure 6-12).

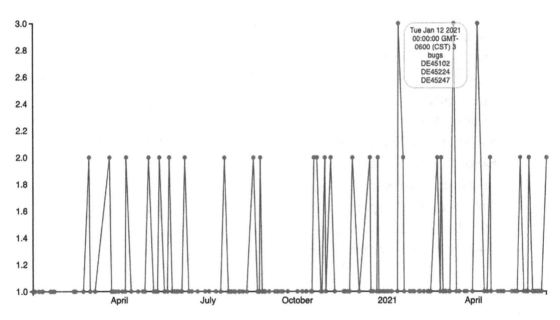

Figure 6-12. *Completed time series with rollover shown*

The complete source code should now look like this:

```
<!DOCTYPE html>
<html>
<meta charset="utf-8">
<head>
<style>
body {
 font: 15px sans-serif;
}
```

179

```css
.trendLine {
 fill: none;
 stroke: #CC0000;
 stroke-width: 1.5px;
}
.axis path{
 fill: none;
 stroke: #000;
 shape-rendering: crispEdges;
}
.tooltip{
        opacity: .75;
        text-align:center;
        font-size:12px;
        width:100px;
        padding:5px;
        border:1px solid #a8b6ba;
        background-color:#fff;
        margin-bottom:5px;
        border-radius: 19px;
        -moz-border-radius: 19px;
        -webkit-border-radius: 19px;
}
</style>
</head>
<body>
        <script src="d3.v3.js"></script>
<script>
var margin = {top: 20, right: 20, bottom: 30, left: 50},
 width = 960 - margin.left - margin.right,
 height = 500 - margin.top - margin.bottom;
var parseDate = d3.time.format("%m/%d/%y").parse;
var xScale = d3.time.scale()
 .range([0, width]);
var yScale = d3.scale.linear()
 .range([height, 0]);
```

```
var xAxis = d3.svg.axis()
        .scale(xScale)
        .orient("bottom");
var yAxis = d3.svg.axis()
        .scale(yScale)
        .orient("left");
var line = d3.svg.line()
 .x(function(d) { return xScale(new Date(d.key)); })
 .y(function(d) { return yScale(d.values.length); });
var tooltip = d3.select("body")
 .append("div")
 .attr("class", "tooltip")
 .attr("id", "tooltip")
 .style("position", "absolute")
 .style("z-index", "10")
 .style("visibility", "hidden");
var svg = d3.select("body").append("svg")
 .attr("width", width + margin.left + margin.right)
 .attr("height", height + margin.top + margin.bottom)
        .append("g")
 .attr("transform", "translate(" + margin.left + "," + margin.top + ")");
d3.csv("https://jonwestfall.com/data/allbugsOrdered.csv", function(error,
data) {
        data.forEach(function(d) {
                d.Date = parseDate(d.Date);
        });
 nested_data = d3.nest()
                .key(function(d) { return d.Date; })
                .entries(data);
        xScale.domain(d3.extent(nested_data, function(d) { return new
        Date(d.key); }));
        yScale.domain(d3.extent(nested_data, function(d) { return
        d.values.length; }));
        svg.append("g")
        .attr("transform", "translate(0," + height + ")")
```

```
        .call(xAxis)
    .attr("class", "axis");
        svg.append("g")
         .call(yAxis)
    .attr("class", "axis");
        svg.append("path")
         .datum(nested_data)
         .attr("d", line)
    .attr("class", "trendLine");
        svg.selectAll("circle")
         .data(nested_data)
         .enter().append("circle")
         .attr("r", 3.5)
         .attr("fill", "red")
         .attr("cx", function(d) { return xScale(new Date(d.key)); })
         .attr("cy", function(d) { return yScale(d.values.
        length);})
                .on("mouseover", function(d){
                        document.getElementById("tooltip").
                        innerHTML = d.key + " " + d.values.length
                        + " bugs<br/>";
                        for(x=0;x<d.values.length;x++){
                                document.getElementById
                                ("tooltip").innerHTML +=
                                d.values[x].ID + "<br/>";
                        }
                        tooltip.style("visibility", "visible");
        })
                .on("mousemove", function(){
                        return tooltip.style("top", (d3.mouse(this)
                        [1] + 25)+"px").style("left", (d3.
                        mouse(this)[0] + 70)+"px");
});
});
```

```
</script>
</body>
</html>
```

Summary

This chapter explored time series plots, both philosophically and in the context of using them to track bug creation over time. We exported the raw bug data from the bug-tracking software of choice and imported it into R to scrub and analyze.

Within R, we looked at different ways we could model and visualize the data, looking at both aggregate and granular details such as how the new bugs contribute to a running total over time or when new bugs are introduced over time. This is especially valuable when we can put context to the dates we are looking at.

We then read the data into D3 and created an interactive time series that allowed us to drill down from the high-level trend data into very granular details around each bug created.

The next chapter explores creating bar charts and how to use them to identify areas of focus and improvement.

Bar Charts

Chapter 6 explored using time series charts to look at defect data over time, and this chapter looks at bar charts, which display ordinal or ranked data relative to a specific data set. They usually consist of an x- and y-axis and have bars or colored rectangles to indicate values of categories.

William Playfair created the bar chart in the first edition of *The Commercial and Political Atlas* in 1786 to show Scotland's import and export data to and from different parts of the world (see Figure 7-1). He created it out of necessity; the other charts in the atlas were time series charts demonstrating hundreds of years' worth of trade data, but for Scotland, there was only one year's worth of data. While using the time series chart, Playfair saw it as an inferior visualization; a compromise with resources on hand because it "does not comprehend any portion of time, and it is much inferior in utility to those that do" (Playfair, 1786, p. 101).

© Tom Barker, Jon Westfall 2022
T. Barker and J. Westfall, *Pro Data Visualization Using R and JavaScript*,
https://doi.org/10.1007/978-1-4842-7202-2_7

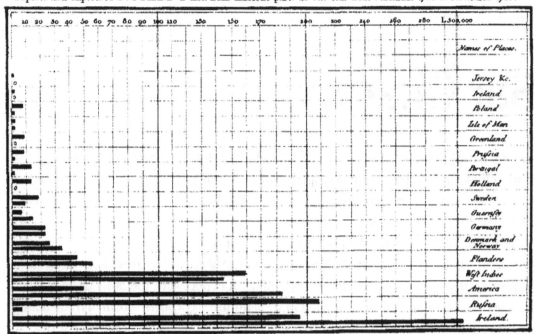

Figure 7-1. *William Playfair's bar chart showing Scotland's import and export data*

Playfair initially thought so little of his invention that he didn't bother to include it in the subsequent second and third editions of the atlas. He went on to envision a different way to show parts of a whole; in doing so, he invented the pie chart for his *Statistical Breviary* published in 1801.

The bar chart is a great way to demonstrate ranked data not only because bars are a clear way to show differences in value but the pattern can also be extended to include more data points by using different types of bar charts such as stacked bar charts and grouped bar charts.

Standard Bar Chart

Let's take data that you are already familiar with—the bugsBySeverity data from the last chapter:

head(bugsBySeverity)

	Blocker	Minor	Moderate
1/11/21	0	1	0
1/12/20	0	1	0
1/12/21	1	2	0
1/13/20	1	0	0
1/17/21	2	0	0
1/18/21	0	0	1

You can create a new list with a sum of each bug type and visualize the totals in a bar chart like so:

```
totalBugsBySeverity <- c(sum(bugsBySeverity[,1]), sum(bugsBySeverity[,2]),
sum(bugsBySeverity[,3]))
barplot(totalBugsBySeverity, main="Total Bugs by Severity")
axis(1, at=1: length(totalBugsBySeverity), lab=c("Blocker", "Critical",
"Minor"))
```

This code produces the chart shown in Figure 7-2.

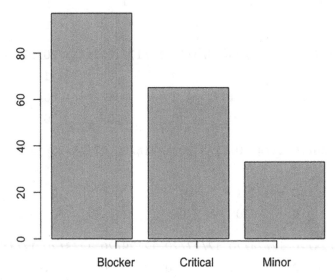

Figure 7-2. *Bar chart of bugs by severity*

Stacked Bar Chart

Stacked bar charts allow us to show subsections or segments within categories. Suppose you use the bugsBySeverity time series data and want to look at the breakdown of the criticality of the new bugs opened each day:

t(bugsBySeverity)

	1/11/21	1/12/20	1/12/21	1/13/20	1/17/21	1/18/21	1/2/21	1/21/20	1/22/20
Blocker	0	0	1	1	2	0	0	1	1
Minor	1	1	2	0	0	0	1	0	0
Moderate	0	0	0	0	0	1	0	0	0

	1/24/20	1/24/21	1/25/20	1/27/21	1/29/21	1/3/20	1/4/20	1/5/20	1/5/21
Blocker	0	0	1	0	0	0	0	0	0
Minor	1	1	0	1	0	1	1	1	1
Moderate	0	0	0	0	1	0	0	0	0

	1/9/20	10/1/20	10/10/20	10/15/20	10/16/20	10/18/20	10/21/20	10/25/20
Blocker	1	0	0	1	0	0	0	1
Minor	0	1	0	0	1	0	1	0
Moderate	0	0	1	0	0	2	1	0

	10/26/20	10/29/20	10/30/20	10/6/20	11/17/20	11/18/20	11/19/20	11/21/20
Blocker	0	1	0	0	0	1	0	0
Minor	0	0	1	1	1	0	1	1
Moderate	1	1	0	0	0	0	0	0

	11/23/20	11/26/20	11/4/20	11/8/20	12/14/20	12/15/20	12/17/20	12/21/20
Blocker	0	2	1	1	1	1	0	1
Minor	1	0	1	0	0	0	1	0
Moderate	0	0	0	0	1	0	0	0

	12/22/20	12/23/20	12/24/20	12/27/20	12/29/20	12/3/20	12/31/20	2/12/21
Blocker	1	0	1	0	0	1	0	1
Minor	0	1	0	0	1	0	1	0
Moderate	1	0	0	1	0	0	0	0

	2/13/21	2/14/20	2/15/20	2/15/21	2/16/20	2/22/21	2/24/20	2/25/21
Blocker	0	1	0	1	1	1	1	0
Minor	0	0	1	0	0	1	0	1
Moderate	1	0	0	0	0	0	0	0

	2/26/21	2/28/21	2/3/21	2/4/21	2/8/21	3/1/20	3/1/21	3/11/21	3/14/21
Blocker	1	1	1	1	1	0	1	2	0
Minor	1	0	0	0	0	0	0	1	1
Moderate	0	0	0	0	0	2	0	0	0

	3/17/21	3/2/20	3/2/21	3/22/20	3/23/21	3/24/20	3/25/21	3/26/20	3/28/20
Blocker	1	1	1	1	0	0	1	0	1
Minor	0	0	0	1	1	0	0	1	0
Moderate	0	0	0	0	0	1	0	0	0

	3/3/21	3/31/20	3/31/21	3/6/21	3/7/20	3/7/21	4/12/21	4/13/20	4/15/21
Blocker	1	0	1	1	0	0	0	0	0
Minor	0	0	0	0	0	0	0	1	0
Moderate	0	1	0	0	1	1	1	0	1

	4/18/21	4/19/21	4/20/20	4/25/20	4/26/21	4/27/20	4/29/21	4/4/20	4/5/21
Blocker	0	0	1	0	1	1	1	0	2
Minor	2	1	0	1	0	0	0	1	1
Moderate	0	0	0	0	0	0	0	0	0

	4/7/20	4/8/20	5/1/20	5/10/20	5/11/21	5/12/20	5/14/21	5/16/21	5/17/20
Blocker	1	1	2	0	1	1	0	1	1
Minor	0	0	0	1	0	1	1	0	0
Moderate	0	1	0	0	0	0	0	0	0

	5/17/21	5/2/21	5/20/20	5/20/21	5/22/20	5/24/21	5/25/20	5/26/21	5/27/20
Blocker	1	1	0	1	2	0	0	1	1
Minor	0	0	0	0	0	1	0	0	0
Moderate	0	0	1	1	0	0	1	0	0

	5/27/21	5/28/20	5/28/21	5/29/21	5/30/20	5/31/20	5/6/20	5/8/20	6/11/20
Blocker	1	0	1	2	1	1	0	1	1
Minor	0	1	0	0	0	0	1	0	0
Moderate	0	0	0	0	0	0	0	0	0

	6/11/21	6/14/20	6/16/21	6/2/21	6/20/20	6/28/20	6/3/20	6/3/21	6/4/20
Blocker	1	1	2	1	1	1	0	1	0
Minor	0	0	0	0	0	0	0	0	1
Moderate	0	0	0	0	0	0	1	0	0

	6/4/21	6/6/21	6/7/20	6/7/21	6/8/21	6/9/21	7/14/20	7/18/20	7/2/20
Blocker	0	1	0	1	0	0	1	2	0
Minor	1	0	1	0	1	1	0	0	1
Moderate	0	0	1	0	0	0	0	0	0

	7/22/20	7/23/20	7/25/20	7/28/20	7/29/20	7/9/20	8/10/20	8/17/20	8/2/20
Blocker	1	0	0	1	0	0	0	0	0
Minor	0	1	0	0	1	1	1	0	1
Moderate	0	0	1	0	0	0	0	2	0

	8/21/20	8/22/20	8/23/20	8/24/20	8/26/20	8/27/20	8/28/20	8/29/20	8/3/20
Blocker	1	0	0	2	1	0	0	1	0
Minor	0	0	1	0	0	1	1	0	1
Moderate	0	1	0	0	0	0	0	0	0

	8/6/20	9/10/20	9/11/20	9/14/20	9/16/20	9/2/20	9/21/20	9/8/20
Blocker	1	1	1	0	0	0	0	0
Minor	0	0	0	0	0	1	1	0
Moderate	0	0	0	1	1	0	0	1

You can represent the following data with a stacked bar chart, as shown in Figure 7-3:

```
barplot(t(bugsBySeverity), col=c("#CCCCCC", "#666666", "#AAAAAA"))
legend("topleft", inset=.01, title="Legend", c("Blocker", "Criticals",
"Minors"), fill=c("#CCCCCC", "#666666", "#AAAAAA"))
```

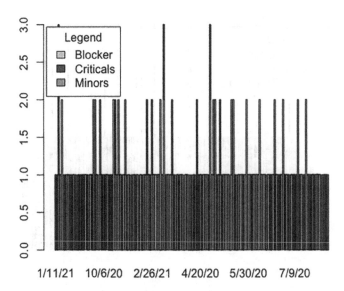

Figure 7-3. *Stacked bar chart of bugs by severity and date. The bars are not all of the same height, since the total number of bugs each day differs*

The total bugs are represented by the full height of the bar, and the colored segments of each bar represent the criticality of the bugs. Stacked bar charts allow us to show nuance in our data, although one may want to reduce the number of dates to get a clearer picture when visualizing.

Grouped Bar Chart

Grouped bar charts allow us to show the same nuance as stacked bar charts, but instead of placing the segments on top of each other, we split them into side-by-side groupings. Figure 7-4 shows that each date on the x-axis has three bars associated with it, one for each criticality category:

```
barplot(t(bugsBySeverity), beside=TRUE, col=c("#CCCCCC", "#666666",
"#AAAAAA"))
legend("topleft", inset=.01, title="Legend", c("Blocker", "Criticals",
"Minors"), fill=c("#CCCCCC", "#666666", "#AAAAAA"))
```

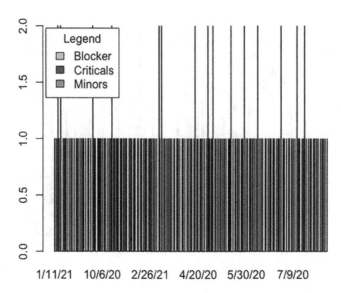

Figure 7-4. *Grouped bar chart of bugs by severity and date*

At a glance, it may appear that Figures 7-3 and 7-4 are identical, due to the density of the data. To avoid this, we can use the following code to reduce the number of data points to just show us five days' worth of data. Try using both snippets to see the changes.

```
barplot(t(bugsBySeverity[1:10,]), col=c("#CCCCCC", "#666666", "#AAAAAA"))
legend("topleft", inset=.01, title="Legend", c("Blocker", "Criticals",
"Minors"), fill=c("#CCCCCC", "#666666", "#AAAAAA"))
```

versus

```
barplot(t(bugsBySeverity[1:10,]), beside=TRUE, col=c("#CCCCCC", "#666666",
"#AAAAAA"))
legend("topleft", inset=.01, title="Legend", c("Blocker", "Criticals",
"Minors"), fill=c("#CCCCCC", "#666666", "#AAAAAA"))
```

Visualizing and Analyzing Production Incidents

If you work on a product that gets used by someone—an end user, a consuming service, or even an internal customer—you most likely have experienced a production incident. Production incidents occur when some part of an application misbehaves for a user in production. It is very much like a bug, but it is a bug that is experienced and reported by your customer.

Just like bugs, production incidents are normal and expected results of software development. There are three main things to think about when talking about incidents:

- **Severity, or how impactful is the error being reported**: There is a big difference between a site outage and a small layout error.

- **Frequency, or how often incidents are occurring or recurring**: If your web app is riddled with issues, your customer experience, your brand, and your regular flow of work are all affected.

- **Duration, or how long individual incidents linger**: The longer they linger, the more customers are affected, and the worse the impact on your brand.

Handling production incidents is a big part of operationalizing your products and maturing your organization. Depending on how severe the incidents are, they can be disruptive to your regular body of work; the team might need to stop everything and work on a fix for the issue. Lesser-priority items can be queued and introduced to the regular body of work alongside regular feature work.

Just as important as handling production incidents is being able to analyze trends in production incidents to identify problem areas. Problem areas are usually features or sections that have frequent issues in production. Once we have identified problem areas, we can do root cause analysis and potentially start to build proactive scaffolding around these areas.

Note   *Proactive scaffolding* is a term I have coined that describes building failovers or additional safety rails to prevent issues in problem areas from recurring. Proactive scaffolding can be anything from detecting when users are close to capacity limits (such as the browser cookie limit or application heap size and correcting before an issue happens) to noting performance issues with third-party assets and intercepting and optimizing them before they are presented to a client.

Another interesting way to handle production incidents is how Heroku used to handle them in the past: putting them up on a timeline along with a visualization of month-over-month uptime and making it publicly available. Heroku's production incident timeline was available at `https://status.heroku.com/`; see Figure 7-5.

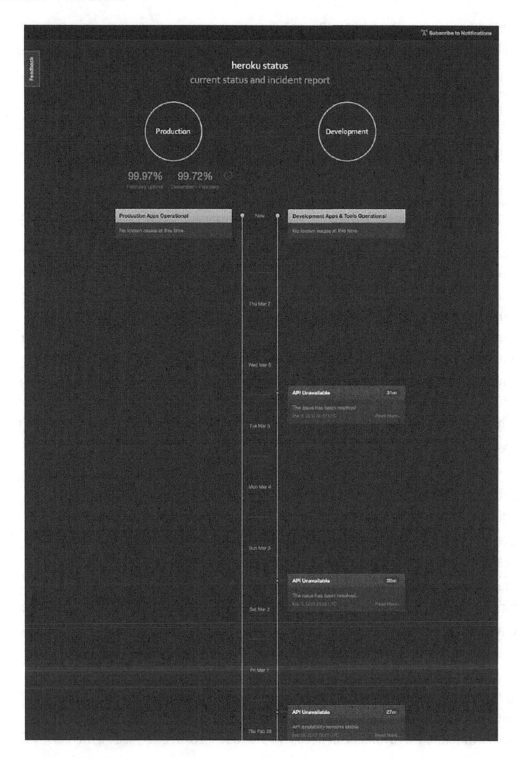

Figure 7-5. *Heroku status page*

GitHub also used to have a great status page that visualizes key metrics around their performance and uptime (see Figure 7-6). Ironically, they've now switched to the timeline approach that Heroku abandoned (see Figure 7-7, from `www.githubstatus.com/history`).

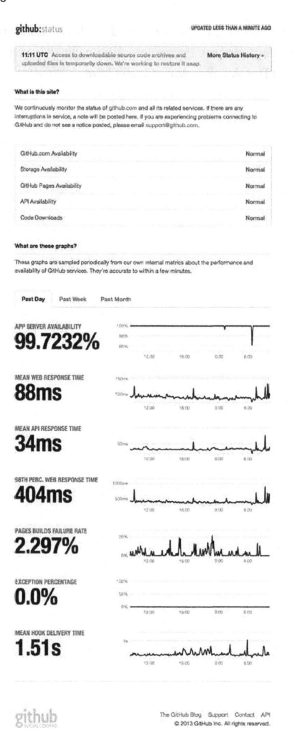

Figure 7-6. *GitHub status page*

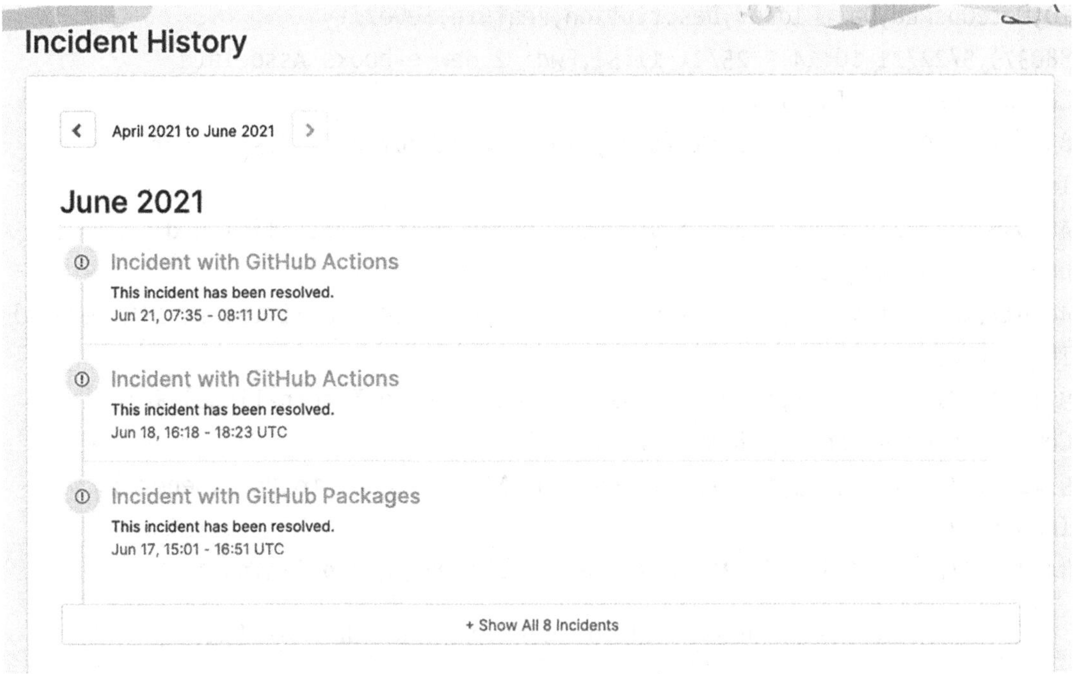

Incident History

< April 2021 to June 2021 >

June 2021

① Incident with GitHub Actions
This incident has been resolved.
Jun 21, 07:35 - 08:11 UTC

① Incident with GitHub Actions
This incident has been resolved.
Jun 18, 16:18 - 18:23 UTC

① Incident with GitHub Packages
This incident has been resolved.
Jun 17, 15:01 - 16:51 UTC

+ Show All 8 Incidents

Figure 7-7. *GitHub's timeline*

For our purposes, this chapter uses bar charts to look at production incidents by feature to start to identify problem areas within our own products.

Plotting Data on a Bar Chart with R

If we want to plot out our production incidents, we must first get an export of the data, just as we needed to do for bugs. Because production incidents are generally single-hit items, companies usually use a range of methods to track them, from ticketing systems such as Jira (`www.atlassian.com/software/jira/overview`) to maintaining a spreadsheet of items, whatever works—as long as we can retrieve the raw data. (Jon has made sample data available here: `http://jonwestfall.com/data/productionincidents.csv`.)

Once we have the raw data, it probably looks something like the following: a comma-separated flat list with columns for an ID, a date stamp, and a description. There also should be a column that lists the feature or section of the application in which the incident occurred.

```
ID,DateOpened,DateClosed,Description,Feature,Severity
880373,5/22/21 10:14,5/25/21 11:52,Fwd: 2 new e-books Associate
Editors,General Inquiry,1
837947,4/29/21 12:35,5/7/21 14:09,Fwd: New Resource to Post,General
Inquiry,2
489036,4/23/21 14:38,4/27/21 9:00,STP ebook editor with finished
book,General Inquiry,1
443617,1/25/21 17:43,1/26/21 8:49,New member - IRC Committee at STP,General
Inquiry,2
911894,1/18/21 10:25,1/20/21 8:51,Fwd: Updates to International Relations
Committee page,General Inquiry,1
974124,1/11/21 14:55,1/12/21 10:55,Fwd: New Resource to Post,General
Inquiry,2
341352,1/2/21 10:51,1/5/21 16:26,New eBooks,eBook Publishing,1
```

Let's read the raw data into R and store it in a variable called prodData:

```
> prodIncidentsFile <- "http://jonwestfall.com/data/productionincidents.csv";
> prodData <- read.table(prodIncidentsFile, sep=",", header=TRUE)
> prodData
      ID    DateOpened      DateClosed  Description
    Feature          Severity
1 880373 5/22/21 10:14  5/25/21 11:52  Fwd: 2 new e-books Associate Editors
    General Inquiry    1
2 837947 4/29/21 12:35   5/7/21 14:09  Fwd: New Resource to Post
    General Inquiry    2
3 489036 4/23/21 14:38   4/27/21 9:00  STP ebook editor with finished book
    General Inquiry    1
4 443617 1/25/21 17:43   1/26/21 8:49  New member - IRC Committee at STP
    General Inquiry    2
5 911894 1/18/21 10:25   1/20/21 8:51  Fwd: Updates to International
                                       Relations Committee page
    General Inquiry    1
6 974124 1/11/21 14:55  1/12/21 10:55  Fwd: New Resource to Post
    General Inquiry    2
7 341352   1/2/21 10:51   1/5/21 16:26  New eBooks
    eBook Publishing  1
```

We want to group them by the Feature column so that we can chart feature totals. To do this, we use the aggregate() function in R. The aggregate() function takes an R object, a list to use as grouping elements, and a function to apply to the grouping elements. So suppose we call the aggregate() function, pass in the ID column as the R object, have it grouped by the Feature column, and have R get the length for each feature grouping:

```
prodIncidentByFeature <- aggregate(prodData$ID, by=list(Feature=prodData$Fe
ature), FUN=length)
```

This code creates an object that looks like the following:

```
> prodIncidentByFeature
            Feature x
1 eBook Publishing 1
2  General Inquiry 6
```

We can then pass this object into the barplot() function to get the chart shown in Figure 7-8.

```
barplot(prodIncidentByFeature$x)
```

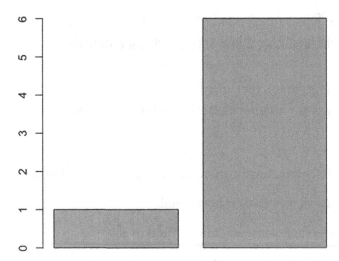

Figure 7-8. *Beginning a bar chart*

This is a nice start and does tell a story, but it's not very descriptive. Besides the fact that the x-axis isn't labeled, the problem areas are obscured by not ordering the results.

Ordering Results

Let's use the order() function to order the results by the total count of each incident by feature:

```
prodIncidentByFeature <- prodIncidentByFeature[order(prodIncidentByFeature$x),]
```

We can then format the bar chart to highlight this ordering by layering the bars horizontally and rotating the text 90 degrees.

To rotate the text, we must change our graphical parameters using the par() function. Updating the graphical parameters has global implications, meaning that any chart that we create after updating inherits the changes, so we need to preserve the current settings and reset them after we create our bar chart. We store our current settings in a variable that we call opar:

```
opar <- par(no.readonly=TRUE)
```

Note   If you are following along in an R command line, the previous line by itself does not generate anything; it just sets graphical parameters.

We then pass new parameters into the par() call. We can use the las parameter to format the axis. The las parameter accepts the following values:

```
par(las=3)
```

- 0 is the default behavior where the text is parallel to the axis.

- 1 explicitly makes the text horizontal.

- 2 makes the text perpendicular to the axis.

- 3 explicitly makes the text vertical.

We then call barplot() again, but this time pass in the parameter horiz=TRUE, to have R draw the bars horizontally instead of vertically:

```
barplot(prodIncidentByFeature$x, xlab="Number of Incidents", names.ar
g=prodIncidentByFeature$Feature, horiz=TRUE, space=1, cex.axis=0.6, cex.
names=0.8, main="Production Incidents by Feature", col= "#CCCCCC")
```

And, finally, we restore the saved settings so that future charts don't inherit this chart's settings:

```
> par(opar)
```

This code produces the visualization shown in Figure 7-9.

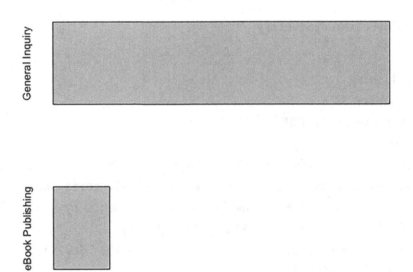

Figure 7-9. *Bar chart of production incidents by feature*

From this chart, you can see that the biggest problem area is the category labeled General Inquiry, followed by eBook Publishing.

Creating a Stacked Bar Chart

How severe are the issues around these features? Let's next create a stacked bar chart to see the breakdown of severity for each production incident. To do that, we must create a table in which we break down our production incidents by feature and by severity. We can use the table() function for this, as we did for bugs in the last chapter:

```
prodIncidentByFeatureBySeverity <- table(factor(prodData$Feature),prodData$
Severity)
```

This code creates a variable formatted as shown in Figure 7-10, with rows representing each feature and columns representing each level of severity:

```
prodIncidentByFeatureBySeverity
```

```
                    1 2
  eBook Publishing 1 0
  General Inquiry  3 3
opar <- par(no.readonly=TRUE)
par(las=3, mar=c(5,5,5,5))
barplot(t(prodIncidentByFeatureBySeverity), xlab="Number of Incidents",
names.arg=rownames(prodIncidentByFeatureBySeverity), horiz=TRUE, space=1,
cex.axis=0.6, cex.names=0.8, main="Production Incidents by Feature",
col=c("#CCCCCC", "#666666", "#AAAAAA", "#333333"))
legend("bottom", inset=.01, title="Legend", c("Sev1", "Sev2"),
fill=c("#CCCCCC", "#666666"))
par(opar)
```

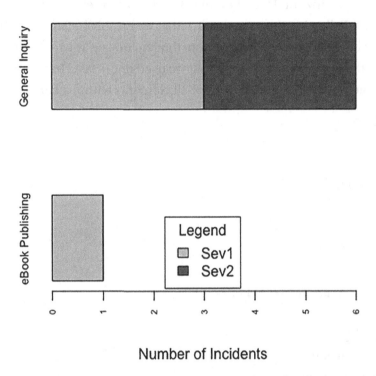

Figure 7-10. *Stacked bar chart of production incidents by feature and by severity*

Interesting! We lost our ordering, but that's because we have a number of new data points to choose from. High-level aggregates are less relevant for this chart; more important is the breakdown of severity.

Bar Charts in D3

So now you know the benefits of having bar charts to aggregate data at a high level and of getting the granular breakdown that stacked bar charts can expose. Let's switch gears and use D3 to see how to create a high-level bar chart that allows us to drill into each bar to see a granular representation of the data at runtime.

We start by creating a bar chart in D3, version 3, and then create a stacked bar chart. When our users mouse over the bar chart, we will overlay the stacked bar chart to show how the data is broken down in real time.

Creating a Vertical Bar Chart

Because we made a horizontal bar chart in D3 back in Chapter 4, we will now make a vertical bar chart. Following the same pattern that we established in previous chapters, we first create a base HTML skeletal structure that includes a link to the D3 version 3 library. We use the same base style rules that we used in the last chapter for body text and axis path and an additional rule to color all elements within a bar class a dark gray.

```
<!DOCTYPE html>
<html>
<head>
<meta charset="utf-8">
<title></title>
<script src="d3.v3.js"></script>
<style type="text/css">
    body {
        font: 15px sans-serif;
    }
    .axis path{
        fill: none;
        stroke: #000;
        shape-rendering: crispEdges;
    }
    .bar {
        fill: #666666;
    }
</style>
</head>
<body></body>
</html>
```

Next, we create the `script` tag to hold all the charting code and the initial set of variables to hold the sizing information: the base height and width, D3 scale objects for the x- and y-coordinate information, an object to hold the margin information, and an adjusted height value that takes the top and bottom margins out of the total height:

```
<script>
var w = 960,
    h = 500,
    x = d3.scale.ordinal().rangeRoundBands([0, w]),
    y = d3.scale.linear().range([0, h]),
    z = d3.scale.ordinal().range(["lightpink", "darkgray", "lightblue"])
    margin = {top: 20, right: 20, bottom: 30, left: 40},
    adjustedHeight = 500 - margin.top - margin.bottom;
</script>
```

We next create the x-axis object. Remember from previous chapters that the axis is not yet drawn, so we need to call it later within the Scalable Vector Graphics (SVG) tag that we will create to draw the axis:

```
var xAxis = d3.svg.axis()
    .scale(x)
    .orient("bottom");
```

Let's draw the SVG container to the page. This will be the parent container for everything else that we will draw to the page.

```
var svg = d3.select("body").append("svg")
    .attr("width", w)
    .attr("height", h)
  .append("g")
```

The next step is to read in the data. We will use the same data source as our R example: the flat file productionIncidents.txt. We can read this in using the d3.csv() function to read in and parse the file. Once the contents of the file are read in, they are stored in the variable data, but if any error occurs, we will store the error details in a variable that we call error.

```
d3.csv("http://jonwestfall.com/data/productionincidents.csv",
function(error, data) {
    }
```

Within the scope of this d3.csv() function is where we will put the majority of our remaining functionality because that functionality depends on having the data proceed.

Let's aggregate the data by feature. To do this, we use the d3.nest() function and set the key to the Feature column:

```
nested_data = d3.nest()
    .key(function(d) { return d.Feature; })
    .entries(data);
```

This code creates an array of objects.

Within this array, each object has a key that lists the feature and an array of objects that list each production incident.

We use this data structure to create the core bar chart. We make a function to do this:

```
function barchart(){
}
```

In this function, we set the transform attribute of the svg element, which sets the coordinates to contain the image that will be drawn. In this case, we constrain it to the margin left and top values:

```
svg.attr("transform", "translate(" + margin.left + "," + margin.top + ")");
```

We also create scale objects for the x- and y-axes. For bar charts, we generally use ordinal scales for the x-axis because they are used for discrete values such as categories. More information about ordinal scales in D3 can be found in the documentation at https://github.com/mbostock/d3/wiki/Ordinal-Scales.

We also create scale objects to map the data to the bounds of the chart:

```
var xScale = d3.scale.ordinal()
    .rangeRoundBands([0, w], .1);
var yScale = d3.scale.linear()
    .range([h, 0]);
xScale.domain(data.map(function(d) { return d.key; }));
yScale.domain([0, d3.max(nested_data, function(d) { return d.values.
length; })]);
```

We next need to draw the bars. We create a selection based on the Cascading Style Sheets (CSS) class that we assign to the bars. We bind the nested_data to the bars, create SVG rectangles for each key value in nested_data, and assign the bar class to each rectangle; we'll define the class style rule soon. We set the x coordinate of each bar to the ordinal scale and set both the y coordinate and the height attribute to the linear scale.

We also add a mouseover event handler and put a call to a function that we will soon create called transitionVisualization(). This function transitions the stacked bar chart that we will make over the bar chart when we mouse over one of the bars.

```
svg.selectAll(".bar")
    .data(nested_data)
    .enter().append("rect")
    .attr("class", "bar")
    .attr("x", function(d) { return xScale(d.key); })
    .attr("width", xScale.rangeBand())
    .attr("y", function(d) { return yScale(d.values.length) - 50; })
    .attr("height", function(d) { return h - yScale(d.values.length); })
    .on("mouseover", function(d){
        transitionVisualization (1)
    })
```

Let's also add in a call to a function that we will create called drawAxes():

```
drawAxes()
```

The complete barchart() function looks like this:

```
function barchart(){
        svg.attr("transform", "translate(" + margin.left + "," + margin.
        top + ")");
        var xScale = d3.scale.ordinal()
            .rangeRoundBands([0, w], .1);
        var yScale = d3.scale.linear()
            .range([h, 0]);
    xScale.domain(nested_data.map(function(d) { return d.key; }));
    yScale.domain([0, d3.max(nested_data, function(d) { return d.values.
    length; })]);
    svg.selectAll(".bar")
        .data(nested_data)
      .enter().append("rect")
        .attr("class", "bar")
```

```
            .attr("x", function(d) { return xScale(d.key); })
            .attr("width", xScale.rangeBand())
            .attr("y", function(d) { return yScale(d.values.length) - 50; })
            .attr("height", function(d) { return h - yScale(d.values.length); })
            .on("mouseover", function(d){
                            transitionVisualization (1)
            })
    drawAxes()
  }
```

Let's create the `drawAxes()` function. We place this function outside the scope of the `d3.csv()` function, out at the root of the `script` tag.

For this chart, let's go with a little more of a minimalist approach and draw only the x-axis. Just like the last chapter, we draw SVG g elements and call the xAxis object:

```
function drawAxes(){
    svg.append("g")
        .attr("class", "x axis")
        .attr("transform", "translate(0," + adjustedHeight + ")")
        .call(xAxis);
}
```

This draws the x-axis that gives the bar chart its category labels.

Creating a Stacked Bar Chart

Now that we have a bar chart, let's create a stacked bar chart. First, let's shape the data. We want an array of objects in which each object represents a feature and has a count of total incidents for each level.

Let's start with a new array called grouped_data:

```
var grouped_data = new Array();
```

Let's iterate through nested_data because nested_data already has taken care of grouping by feature:

```
nested_data.forEach(function (d) {
}
```

Within each pass through nested_data, we create a temporary object and iterate through each incident within the values array:

```
tempObj = {"Feature": d.key, "Sev1":0, "Sev2":0, "Sev3":0, "Sev4":0};
    d.values.forEach(function(e){
    }
```

Within each iteration in the values array, we test the severity of the current incident and increment the appropriate property of the temporary object:

```
if(e.Severity == 1)
    tempObj.Sev1++;
else if(e.Severity == 2)
    tempObj.Sev2++
else if(e.Severity == 3)
    tempObj.Sev3++;
else if(e.Severity == 4)
    tempObj.Sev4++;
```

The complete code to create the grouped_data array looks like the following:

```
nested_data.forEach(function (d) {
    tempObj = {"Feature": d.key, "Sev1":0, "Sev2":0, "Sev3":0, "Sev4":0};
    d.values.forEach(function(e){
        if(e.Severity == 1)
            tempObj.Sev1++;
        else if(e.Severity == 2)
            tempObj.Sev2++
        else if(e.Severity == 3)
            tempObj.Sev3++;
        else if(e.Severity == 4)
            tempObj.Sev4++;
    })
    grouped_data[grouped_data.length] = tempObj
});
```

Perfect! Next, we create a function in which we draw the stacked bar chart within the scope of the d3.csv() function:

```
function stackedBarChart(){
}
```

Here's where it gets interesting. Using the d3.layout.stack() function, we transpose our data so that we have an array in which each index represents one of the levels of severity and contains an object for each feature that has a count of each incident for the respective level of severity:

```
var sevStatus = d3.layout.stack()(["Sev1", "Sev2", "Sev3", "Sev4"].
map(function(sevs)
    {
        return grouped_data.map(function(d) {
        return {x: d.Feature, y: +d[sevs]};
    });
  }));
```

We next use sevStatus to create domain maps for the x and y values of the bar segments that we will draw:

```
x.domain(sevStatus[0].map(function(d) { return d.x; }));
y.domain([0, d3.max(sevStatus[sevStatus.length - 1], function(d) { return
d.y0 + d.y; })]);
```

Next, we draw SVG g elements for each index in the sevStatus array. They serve as containers in which we draw the bars. We bind sevStatus to these grouping elements and set the fill attribute to return one of the colors from the array of colors.

```
var sevs = svg.selectAll("g.sevs")
    .data(sevStatus)
    .enter().append("g")
    .attr("class", "sevs")
    .style("fill", function(d, i) { return z(i); });
```

Finally, we draw the bars within the groupings that we just created. We bind a generic function to the data attribute of the bars that just passes through whatever data is passed to it; this inherits from the SVG groupings.

We draw the bars with the opacity set to 0, so the bars are not initially visible. We also attach mouseover and mouseout event handlers, to call transitionVisualization()—passing 1 when the mouseover event is fired and 0 when the mouseout event is fired (we will flesh out the functionality of transitionVisualization() very soon).

```
var rect = sevs.selectAll("rect")
    .data(function(data){ return data; })
    .enter().append("svg:rect")
    .attr("x", function(d) { return x(d.x) + 13; })
    .attr("y", function(d) { return -y(d.y0) - y(d.y) + adjustedHeight; })
    .attr("class", "groupedBar")
    .attr("opacity", 0)
    .attr("height", function(d) { return y(d.y) ; })
    .attr("width", x.rangeBand() - 20)
    .on("mouseover", function(d){
        transitionVisualization (1)
    })
    .on("mouseout", function(d){
    transitionVisualization (0)
    });
```

The complete stacked bar chart code should look like the following

```
function groupedBarChart(){
    var sevStatus = d3.layout.stack()(["Sev1", "Sev2", "Sev3", "Sev4"].
    map(function(sevs)
    {
        return grouped_data.map(function(d) {
        return {x: d.Feature, y: +d[sevs]};
    });
}));
    x.domain(sevStatus[0].map(function(d) { return d.x; }));
    y.domain([0, d3.max(sevStatus[sevStatus.length - 1], function(d) {
    return d.y0 + d.y; })]);
// Add a group for each sev category.
    var sevs = svg.selectAll("g.sevs")
        .data(sevStatus)
```

```
        .enter().append("g")
        .attr("class", "sevs")
        .style("fill", function(d, i) { return z(i); })
        .style("stroke", function(d, i) { return d3.rgb(z(i)).darker(); });
    var rect = sevs.selectAll("rect")
        . data(function(data){ return data; })
        .enter().append("svg:rect")
        .attr("x", function(d) { return x(d.x) + 13; })
        .attr("y", function(d) { return -y(d.y0) - y(d.y) +
         adjustedHeight; })
         .attr("class", "groupedBar")
        .attr("opacity", 0)
        .attr("height", function(d) { return y(d.y) ; })
        .attr("width", x.rangeBand() - 20)
        .on("mouseover", function(d){
            transitionVisualization (1)
        })
        .on("mouseout", function(d){
        transitionVisualization (0)
        });
    }
```

Creating an Overlaid Visualization

But we're not quite done yet. We've been referencing this transitionVisualization()
function, but we haven't yet defined it. Let's take care of that right now. Remember
how we've been using it: when a user mouses over a bar in our bar chart, we call
transitionVisualization() and pass in a 1. When a user mouses over a bar in our stacked
bar chart, we also call transitionVisualization() and pass in a 1. But when a user mouses
off a bar in the stacked bar chart, we call transitionVisualization() and pass in a 0.

So the parameter that we pass in sets the opacity of our stacked bar chart. Because we
initially draw the stacked bar chart with the opacity at 0, we only ever see it when a user rolls
over a bar in the bar chart, and it gets hidden again when the user rolls off of the bar.

To create this effect, we use a D3 transition. Transitions are much like tweens in
other languages such as ActionScript 3. We create a D3 selection (in this case, we can

select all elements of class groupedBar), call transition(), and set the attributes of that selection that we want to change:

```
function transitionVisualization(vis){
    var rect = svg.selectAll(".groupedBar")
    .transition()
    .attr("opacity", vis)
}
```

We've now got our entire visualization, as can be seen in Figure 7-11.

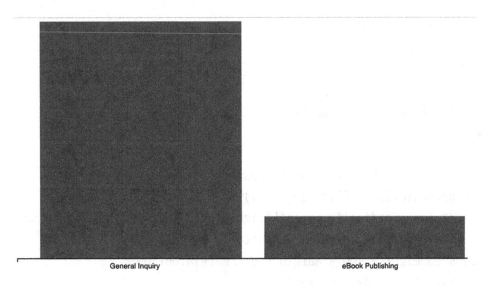

Figure 7-11. *Stacked bar chart of production incidents by feature and by severity*

The completed code looks like the following, and although it's hard to demonstrate this functionality via a printed medium, you can see the working model on Jon's website (available at https://jonwestfall.com/d3/ch7.d3.example.htm) or put the code onto a local web server and run it yourself:

```
<!DOCTYPE html>
<html>
  <head>
        <meta charset="utf-8">
    <title></title>
        <script src="d3.v3.js"></script>
```

```
        <style type="text/css">
        body {
          font: 15px sans-serif;
        }
        .axis path{
          fill: none;
          stroke: #000;
          shape-rendering: crispEdges;
        }
        .bar {
          fill: #666666;
        }
    </style>   </head>
  <body>
    <script type="text/javascript">
var w = 960,
    h = 500,
    x = d3.scale.ordinal().rangeRoundBands([0, w]),
    y = d3.scale.linear().range([0,h]),
    z = d3.scale.ordinal().range(["lightpink", "darkgray", "lightblue"])
    margin = {top: 20, right: 20, bottom: 30, left: 40},
    adjustedHeight = 500 - margin.top - margin.bottom;
        var xAxis = d3.svg.axis()
            .scale(x)
            .orient("bottom");
        var svg = d3.select("body").append("svg")
            .attr("width", w)
            .attr("height", h)
          .append("g")
        function drawAxes(){
          svg.append("g")
              .attr("class", "x axis")
              .attr("transform", "translate(0," + adjustedHeight + ")")
              .call(xAxis);
        }
```

```
      function transitionVisuaization(vis){
              var rect = svg.selectAll(".groupedBar")
              .transition()
              .attr("opacity", vis)
      }
    d3.csv("https://jonwestfall.com/data/productionincidents.csv",
    function(error, data) {
        nested_data = d3.nest()
                     .key(function(d) { return d.Feature; })
                     .entries(data);
            var grouped_data = new Array();
            //for stacked bar chart
            nested_data.forEach(function (d) {
                    tempObj = {"Feature": d.key, "Sev1":0, "Sev2":0,
                    "Sev3":0, "Sev4":0};
                    d.values.forEach(function(e){
                            if(e.Severity == 1)
                            tempObj.Sev1++;
                            else if(e.Severity == 2)
                            tempObj.Sev2++
                            else if(e.Severity == 3)
                            tempObj.Sev3++;
                            else if(e.Severity == 4)
                            tempObj.Sev4++;
                    })
                    grouped_data[grouped_data.length] = tempObj
            });
function stackedBarChart(){
  var sevStatus = d3.layout.stack()(["Sev1", "Sev2", "Sev3", "Sev4"].
  map(function(sevs) {
    return grouped_data.map(function(d) {
      return {x: d.Feature, y: +d[sevs]};
    });
  }));
```

```
x.domain(sevStatus[0].map(function(d) { return d.x; }));
y.domain([0, d3.max(sevStatus[sevStatus.length - 1], function(d) { return
d.y0 + d.y; })]);
// Add a group for each sev category.
var sevs = svg.selectAll("g.sevs")
    .data(sevStatus)
  .enter().append("g")
    .attr("class", "sevs")
    .style("fill", function(d, i) { return z(i); });
var rect = sevs.selectAll("rect")
    .data(function(data){ return data; })
  .enter().append("svg:rect")
    .attr("x", function(d) { return x(d.x) + 13; })
    .attr("y", function(d) { return -y(d.y0) - y(d.y) + adjustedHeight; })
        .attr("class", "groupedBar")
        .attr("opacity", 0)
    .attr("height", function(d) { return y(d.y) ; })
    .attr("width", x.rangeBand() - 20)
        .on("mouseover", function(d){
                transitionVisuaization(1)
        })
        .on("mouseout", function(d){
                transitionVisuaization(0)
        });
}
function barchart(){
        svg.attr("transform", "translate(" + margin.left + "," + margin.
        top + ")");
        var xScale = d3.scale.ordinal()
            .rangeRoundBands([0, w], .1);
        var yScale = d3.scale.linear()
            .range([h, 0]);
    xScale.domain(nested_data.map(function(d) { return d.key; }));
    yScale.domain([0, d3.max(nested_data, function(d) { return d.values.
    length; })]);
```

```
    svg.selectAll(".bar")
        .data(nested_data)
      .enter().append("rect")
        .attr("class", "bar")
        .attr("x", function(d) { return xScale(d.key); })
        .attr("width", xScale.rangeBand())
        .attr("y", function(d) { return yScale(d.values.length) - 50; })
        .attr("height", function(d) { return h - yScale(d.values.length); })
        .on("mouseover", function(d){
                        transitionVisuaization(1)
        })
      stackedBarChart()
    drawAxes()
  }
  barchart();
});
    </script>
  </body>
</html>
```

Summary

This chapter looked at using bar charts to display ranked data in the context of production incidents. Because production incidents are essentially direct feedback from your user base around how your product is misbehaving or failing, managing production incidents is a critical piece of any mature engineering organization.

Managing production incidents isn't simply about responding to issues as they arise, however; it is also about analyzing the data around your incidents: what areas of your application are breaking frequently, what unexpected patterns of use you see in production that could cause these recurring issues, how to build proactive scaffolding to prevent these and future issues. All these are questions you can answer only by fully understanding your product and your data. In this chapter, you took your first step toward that greater understanding.

Correlation Analysis with Scatter Plots

In the last chapter, you looked at using bar charts to analyze production incidents. You saw that bar charts are great for displaying the differences in a ranked data set, and you used this idea to identify areas in which issues recurred. You also used stacked bar charts to see the granular breakdown in the severity of production incidents.

This chapter looks at correlation analysis with scatter plots. Scatter plots are charts that plot two independent data sets on their own axes, displayed as points on a Cartesian grid (x and y coordinates). As you'll see, scatter plots are used to try and identify relationships between the two data points.

Note   Michael Friendly and Daniel Denis have published a thoughtful and thoroughly researched dissertation on the history of scatter plots, originally published by the *Journal of the History of the Behavioral Sciences*, Vol. 41, in 2005 and available on Friendly's website at `www.datavis.ca/papers/friendly-scat.pdf`. This article is absolutely recommended reading because it tries to trace back the very first recorded scatter plots and the first time a chart was called a scatter plot and very deftly delineates the difference between a scatter plot and a time series (in other words, all time series are scatter plots with time as an axis while not all scatter plots are time series!).

© Tom Barker, Jon Westfall 2022
T. Barker and J. Westfall, *Pro Data Visualization Using R and JavaScript*,
https://doi.org/10.1007/978-1-4842-7202-2_8

Finding Relationships in Data

The pattern, or lack of a pattern, that the points form on a scatter plot indicates the relationship. At a very high level, relationships can be

- Positive correlation, in which one variable increases as the other increases. This is demonstrated by the dots forming a line trending diagonally upward from left to right (see Figure 8-1).

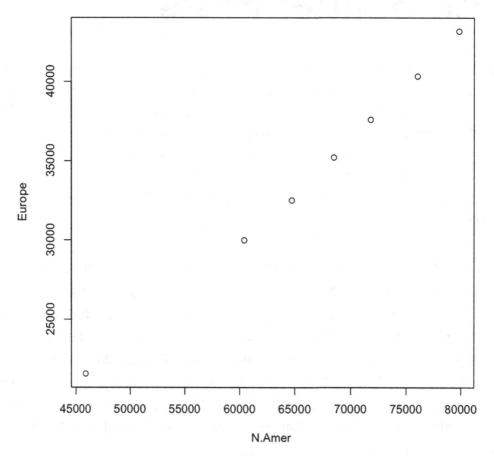

Figure 8-1. *Scatter plot showing positive correlation between total phones in North America and Europe*

- Negative correlation, in which one variable increases as the other decreases. This is demonstrated by the dots forming a line trending downward from left to right (see Figure 8-2).

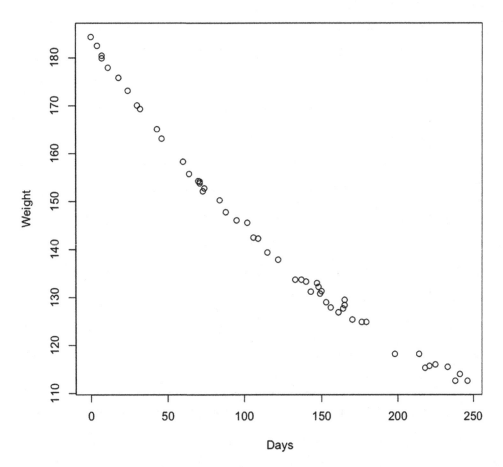

Figure 8-2. *Scatter plot showing negative correlation between body weight and time passing (for a person on a diet)*

- No correlation, demonstrated (or not) by a scatter plot that has no discernible trend line (see Figure 8-3).

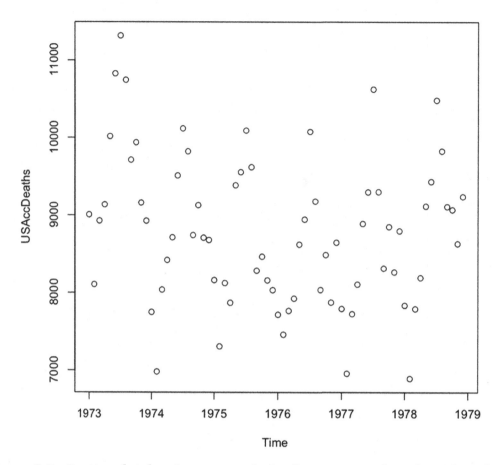

Figure 8-3. *Scatter plot showing no correlation between number of accidental deaths in the United States over years*

Of course, simply identifying correlation between two data points or data sets does not imply that there is direct cause in the relationship—hence the convention that correlation does not imply causation. For example, see the negative correlation chart in Figure 8-2. If we were to assume direct causation between the two axes—weight and number of days—we would be assuming that the passing of time caused body weight to decrease.

Although scatter plots are great for analyzing the relationship between two sets of data, there is a related pattern that can be used to introduce a third set of data as well. This visualization is called a bubble chart, and it uses the radius of the points in a scatter plot to expose the third dimension of data.

See Figure 8-4 for a bubble chart that shows the correlation in length of tooth growth in guinea pigs and doses of vitamin C administered. The third data point is the method of delivery: either by vitamin supplement or by orange juice. It is added as the radius of each point in the graphic; the larger circle is the vitamin supplement, and the smaller circle is orange juice.

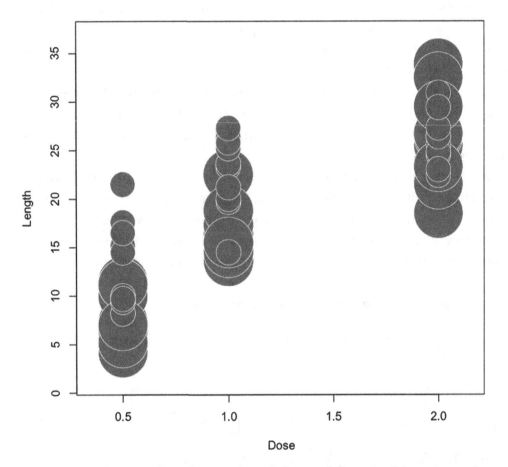

Figure 8-4. *Correlation of tooth growth and doses of vitamin C in guinea pigs, both by vitamin supplement and by orange juice*

For our purposes in this chapter, we will use scatter plots and bubble charts to look at the implied relationship that team velocity has with our other areas of focus, in effect doing correlation analysis on team dynamics. We will compare things like team size and velocity, velocity and production incidents, and so on.

Introductory Concepts of Agile Development

Let's start by introducing some preliminary concepts of Agile development. If you are already versed in Agile, this section will be a bit of a review. There are many flavors of Agile development, but the high-level concepts that most have in common are the ideas of time boxing a body of work. Time boxing enables the team to focus on one thing and finish it, allowing the stakeholders to quickly give feedback on what was completed. This short feedback loop allows for teams and stakeholders to pivot, or react and change direction as requirements and even industries change.

This span of time that the team works on the body of work—whether it is one week, three weeks, or what have you—is called a sprint. At the end of a sprint, the team doing the work should have releasable code, though releasing after each sprint is not a requirement.

Sprints begin with a planning session in which teams define the body of work, and sprints end with a review session in which the team goes over the body of work completed. Periodically during a sprint, the team grooms new work to complete; it defines the work in user stories that list acceptance criteria. It is these user stories that get prioritized and committed to in the planning sessions held at the beginning of each sprint.

See Figure 8-5 for a high-level workflow of this process.

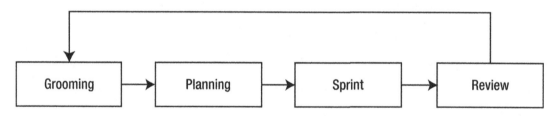

Figure 8-5. *High-level workflow for Agile development*

User stories have story points associated with them. Story points are estimates of the level of complexity for the story and are usually a numeric value. As teams complete sprints, they begin to form a consistent velocity. Velocity is the average amount of story points that a team will complete in a sprint.

Velocity is important because you use it to estimate how much your team can complete at the start of each sprint and to project out how much of your backlog of work the team may be able to complete from your roadmap over the course of the year.

There are a number of tools available to manage Agile projects, such as Rally (`www.rallydev.com/`) or Greenhopper from Atlassian (`www.atlassian.com/software/greenhopper/overview`), the same company that makes Jira and Confluence. Whatever tool you use should provide the ability to export your data, including user point counts for each sprint.

Correlation Analysis

To begin the analysis, let's export a totaled sum of story points for each sprint along with the team name. We should compile all these data points into a single file that we will name `teamvelocity.txt`. Our file should look something like the following, which shows data for the 12.1 and 12.2 sprints for the teams named Red and Gold (arbitrary names for teams that are working on the same product just with different bodies of work):

```
Sprint,TotalPoints,Team
12.1,25,Gold
12.1,63,Red
12.2,54,Red
...
```

Let's add an additional column in there to represent the total team members on each team for each sprint. The data should now look like so:

```
Sprint,TotalPoints,TotalDevs,Team
12.1,25,6,Gold
12.1,63,10,Red
12.2,54,9,Red
...
```

We have also made this sample data set available, with more points, here: `https://jonwestfall.com/data/teamvelocity.txt`.

Excellent! Let's now read this into R, changing the path in the first line to be where you have placed it:

```
tvFile <- "/Applications/MAMP/htdocs/teamvelocity.txt"
teamvelocity <- read.table(tvFile, sep=",", header=TRUE)
```

Creating a Scatter Plot

Now create a scatter plot using the plot() function to compare the total points that the teams completed in each sprint against how many members were on the team for each sprint. We pass teamvelocity$TotalPoints and teamvelocity$TotalDevs as the first two parameters, set the type to p, and give meaningful labels for the axes:

```
plot(teamvelocity$TotalPoints,teamvelocity$TotalDevs, type="p", ylab="Team
Members", xlab="Velocity", bg="#CCCCCC", pch=21)
```

This creates the scatter plot that we can see in Figure 8-6; we can see that as we add more members to a team, the number of story points that they can complete in an iteration, or sprint, also increases.

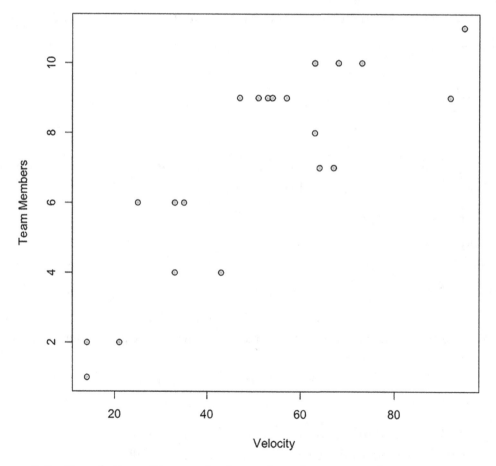

Figure 8-6. _Correlation of team velocity and total team members_

Creating a Bubble Chart

If we want a greater insight into the data that we have so far, for example, to show which points belong to which team, we could visualize that information with a bubble chart. We can create bubble charts using the `symbols()` function. We pass in `TotalPoints` and `TotalDevs` into `symbols()`, just as we did for `plot()`, but we also pass in the `Team` column into a parameter named `circles`. This specifies the radius of the circle to draw on the chart. Because for our example `Team` is a string, R converts it to a factor. We also set the color of the circle with the `bg` parameter and the stroke color of the circle with the `fg` parameter.

```
symbols(teamvelocity$TotalPoints, teamvelocity$TotalDevs, circles=as.
factor(teamvelocity$Team), inches=0.35, fg="#000000", bg="#CCCCCC",
ylab="Team Members", xlab="Velocity")
```

The previous R code should produce a bubble chart that looks like Figure 8-7.

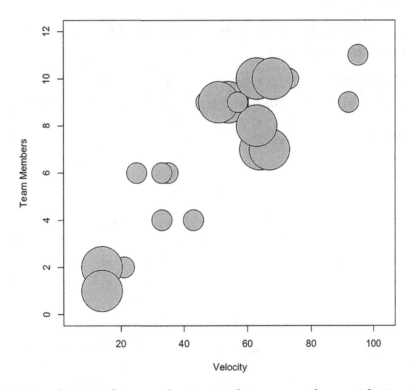

Figure 8-7. *Correlation of team velocity, total team members, with size of bubble indicating team*

Visualizing Bugs

The bubble chart shown in Figure 8-7 is of limited use, mainly because the team breakdown is not really a relevant data point. Let's take the `teamvelocity.txt` file and begin to layer in more information. We already discussed tracking bug data back in Chapter 6; now let's use our bug-tracking software and add in two new bug-related data points: the total bugs in each team's backlog at the end of each sprint and how many bugs were opened within each sprint. We'll name the columns for these new data points `BugBacklog` and `BugsOpened`, respectively.

The updated file should look something like this:

```
Sprint,TotalPoints,TotalDevs,Team,BugBacklog,BugsOpened
12.1,25,6,Gold,125,10
12.2,42,8,Gold,135,30
12.3,45,8,Gold,150,25
```

Let's next create a scatter plot with this new data. We'll first compare velocity against bugs opened during each iteration:

```
plot(teamvelocity$TotalPoints,teamvelocity$BugsOpened, type="p",
xlab="Velocity", ylab="Bugs Opened During Sprint", bg="#CCCCCC", pch=21)
```

This creates the scatter plot shown in Figure 8-8.

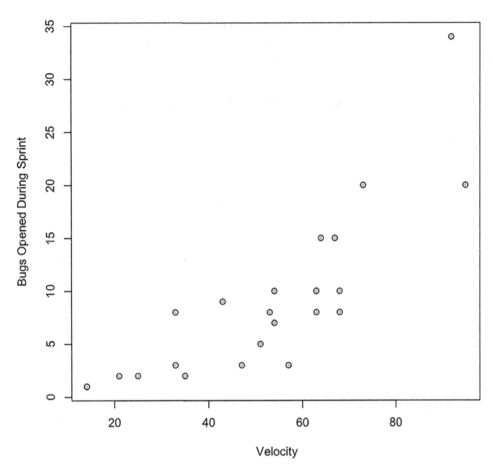

Figure 8-8. *Correlation of team velocity and bugs opened*

Now this is very interesting. There is a positive correlation between having more people on a team and getting more done (or at least getting more complex work done), and the more story points that are completed, the more bugs are generated. So an increase in complexity correlates to an increase in the number of bugs created in a given sprint. At least that seems to be implied by my data.

Let's reflect this new data point in the existing bubble chart; instead of sizing circles by team, we size them by bugs opened:

```
symbols(teamvelocity$TotalPoints, teamvelocity$TotalDevs, circles=
teamvelocity$BugsOpened, inches=0.35, fg="#000000", bg="#CCCCCC",
ylab="Team Members", xlab="Velocity", main = "Velocity by Team Size by Bugs
Opened")
```

This code produces the bubble chart shown in Figure 8-9; you see that the sizing of the bubbles follows the existing pattern of positive correlation, in that the bubbles get larger as both the number of team members and the team velocity increases.

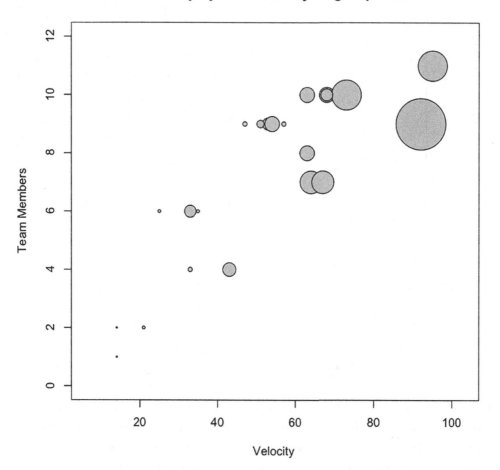

Figure 8-9. *Correlation of team velocity and team size, where circle size indicates bugs opened*

Let's next create a scatter plot to look at the total bug backlog after each sprint:

```
plot(teamvelocity$TotalPoints,teamvelocity$BugBacklog, type="p",
xlab="Velocity", ylab="Total Bug Backlog", bg="#CCCCCC", pch=21)
```

This code produces the chart shown in Figure 8-10.

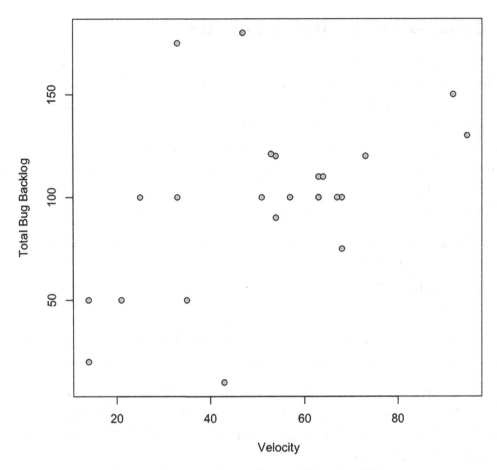

Figure 8-10. *Correlation of team velocity by total bug backlog*

This figure shows that no correlation exists. This could be because of any number of reasons: maybe the team has been fixing bugs during the sprint, or maybe they are closing all the bugs opened during the course of the iteration. Determining the root cause is beyond the scope of the scatter plot, but we can tell that while the bugs being opened and the level of complexity increases, the total bug backlog does not increase.

Visualizing Production Incidents

Let's next layer in another data point into the file; we'll add a column for production incidents opened against the work done during the sprint. To be very specific, when a body of work in a sprint is completed, it is released to production, and a release number is generally associated with that release. This last data point we discuss is concerned with tracking issues in production against the release for a given iteration. Not issues that came in during the iteration; issues that came in once the work done in the iteration was pushed to production.

Now let's add in the last column, named `ProductionIncidents`:

```
Sprint,TotalPoints,TotalDevs,Team,BugBacklog,BugsOpened,ProductionIncidents
12.1,25,6,Gold,125,10,1
12.2,42,8,Gold,135,30,3
12.3,45,8,Gold,150,25,2
```

Great! Let's next create a new bubble chart with this data, comparing total story points completed, bugs opened each iteration, and production incidents per release:

```
symbols(teamvelocity$TotalPoints, teamvelocity$BugsOpened, circles=team
velocity$ProductionIncidents, inches=0.35, fg="#000000", bg="#CCCCCC",
ylab="Bugs Opened", xlab="Velocity", main = "Velocity by Bugs Opened by
Production Incidents Opened")
```

This code creates the chart shown in Figure 8-11.

Velocity by Bugs Opened by Production Incidents Opened

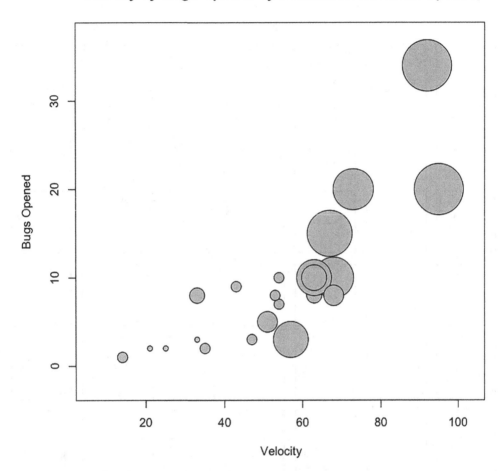

Figure 8-11. *Correlation of team velocity and bugs opened, where the size of the circle indicates the number of production incidents*

From this chart, you can see that, at least according to our sample data, there exists a positive correlation between total story points completed, bugs opened, and production incidents opened for a given sprint.

Finally, now that all the data is layered into the flat file, we can create a scatter plot matrix. This is a matrix of all the columns compared with each other with scatter plots. We can use the scatter plot matrix to look at all the data at once and quickly pick out any correlation patterns that may exist in the data set. We can create a scatter plot matrix with just the plot() function or with the pairs() function in the graphics package:

```
plot(teamvelocity)
pairs(teamvelocity)
```

Either one produces the chart shown in Figure 8-12.

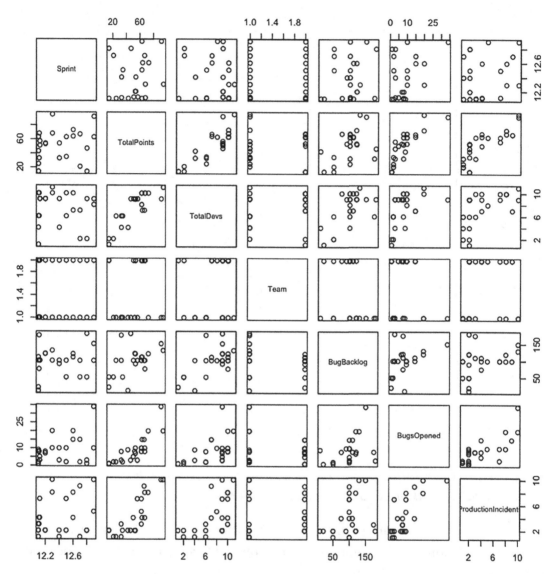

Figure 8-12. *Scatter plot matrix of our complete data set*

In Figure 8-12, each row represents one of the columns in the data frame, and each scatter plot represents the intersection of those columns. When you scan over each scatter plot in the matrix, you can clearly see the correlation patterns in the combinations already covered this chapter. While an effective visualization, by looking at so many variables at the same time, the eye can easily get fatigued. It's important to think about how just because you can put everything in one figure, you might not want to. You may consider subsetting your data to just certain columns of interest, making a figure like this easier to walk through.

Interactive Scatter Plots in D3

So far in this chapter, we've been creating different scatter plots to represent the data combinations that we wanted to look at. But what if we want to create a scatter plot that allows us to select the data points on which the axes were based? With D3, we can do just that!

Adding the Base HTML and JavaScript

Let's start with the base HTML structure that has d3.js included as well as the base CSS:

```
<!DOCTYPE html>
<html>
  <head>
          <meta charset="utf-8">
    <title></title>
<style>
body {
  font: 15px sans-serif;
}
.axis path{
  fill: none;
  stroke: #000;
  shape-rendering: crispEdges;
}
.dot {
  stroke: #000;
}
</style>
</head>
<body>
```

```
<script src="d3.v3.js"></script>
</body>
</html>
```

Let's next add in the script tag to hold the chart. Just like the previous D3 examples, include the starting variables, margin, x and y range objects, and x- and y-axis objects:

```
<script>
var margin = {top: 20, right: 20, bottom: 30, left: 40},
    width = 960 - margin.left - margin.right,
    height = 500 - margin.top - margin.bottom;
var x = d3.scale.linear()
    .range([0, width]);
var y = d3.scale.linear()
    .range([height, 0]);
var xAxis = d3.svg.axis()
    .scale(x)
    .orient("bottom");
var yAxis = d3.svg.axis()
    .scale(y)
    .orient("left");
</script>
```

Let's also create the SVG tag on the page as in the previous examples:

```
var svg = d3.select("body").append("svg")
    .attr("width", width + margin.left + margin.right)
    .attr("height", height + margin.top + margin.bottom)
  .append("g")
    .attr("transform", "translate(" + margin.left + "," + margin.top + ")");
```

Loading the Data

Now we need to load in the data using the d3.csv() function. In all previous D3 examples, most of the work was done in the scope of the callback function, but for this example, we need to expose our functionality publicly so we can change the data points via form select elements. Yet we still need to drive the initial functionality from the callback function because that's when we will have our data, so we will set up our callback function to call stubbed out public functions.

We set a public variable that we call chartData to the data returned from the flat file and call two functions called removeDots() and setChartDots():

```
d3.csv("teamvelocity.txt", function(error, data) {
        chartData = data;
        removeDots()
        setChartDots("TotalDevs", "TotalPoints")
});
```

Notice that we passed in "TotalDevs" and "TotalPoints" to the setChartDots() function. This is to prime the pump because they will be the initial data points we show when the page loads.

Adding Interactive Functionality

Now we need to actually create the things we stubbed out. First, let's create the variable chartData at the root of the script tag where we set the other variables:

```
var margin = {top: 20, right: 20, bottom: 30, left: 40},
    width = 960 - margin.left - margin.right,
    height = 500 - margin.top - margin.bottom,
    chartData;
```

Next, we create the removeDots() function, which selects any circles or axes on the page and removes them:

```
function removeDots(){
    svg.selectAll("circle")
        .transition()
            .duration(0)
            .remove()
    svg.selectAll(".axis")
            .transition()
            .duration(0)
            .remove()
}
```

And, finally, we create the `setChartDots()` functionality. The function accepts two parameters: `xval` and `yval`. Because we want to make sure that the D3 transitions are finished running and they have a 250-millisecond default runtime, even when we set the duration to 0, we will wrap the contents of the function in a `setTimeout()` call, so we wait 300 milliseconds before starting to draw our chart. If we don't do this, we could enter into a race condition in which we are drawing to screen as the transition is removing from the screen.

```
function setChartDots(xval, yval){
        setTimeout(function() {
        }, 300);
  }
```

Within the function, we set the domains of the x and y scale objects using the `xval` and `yval` parameters. These parameters correspond to the column names of the data points that we will be charting:

```
x.domain(d3.extent(chartData, function(d) { return d[xval];}));
y.domain(d3.extent(chartData, function(d) { return d[yval];}));
```

Next, we draw the circles to the screen, using the global `chartData` variable to feed it and the passed-in columnal data as the x and y coordinates of the circles. We also grow the axes in this function, so that we redraw the values each time an axis is changed.

```
svg.selectAll(".dot")
    .data(chartData)
    .enter().append("circle")
    .attr("class", "dot")
    .attr("r", 3)
    .attr("cx", function(d) { return x(d[xval]);})
    .attr("cy", function(d) { return y(d[yval]);})
    .style("fill", "#CCCCCC");
svg.append("g")
    .attr("class", "axis")
    .attr("transform", "translate(0," + height + ")")
    .call(xAxis)
  svg.append("g")
    .attr("class", "axis")
    .call(yAxis)
```

The complete function should look like the following:

```
function setChartDots(xval, yval){
      setTimeout(function() {
        x.domain(d3.extent(chartData, function(d) { return d[xval];}));
        y.domain(d3.extent(chartData, function(d) { return d[yval];}));
        svg.selectAll(".dot")
            .data(chartData)
          .enter().append("circle")
            .attr("class", "dot")
            .attr("r", 3)
            .attr("cx", function(d) { return x(d[xval]);})
            .attr("cy", function(d) { return y(d[yval]);})
            .style("fill", "#CCCCCC");
            svg.append("g")
                .attr("class", "axis")
                .attr("transform", "translate(0," + height + ")")
                .call(xAxis)
            svg.append("g")
                .attr("class", "axis")
                .call(yAxis)
      }, 300);
}
```

Excellent!

Adding Form Fields

Let's next add in the form fields. We'll add two select elements, where each option corresponds to a column in the flat file. The elements call a JavaScript function, getFormData(), that we will define shortly:

```
<form>
      Y-Axis:
      <select id="yval" onChange="getFormData()">
              <option value="TotalPoints">Total Points</option>
              <option value="TotalDevs">Total Devs</option>
```

```
            <option value="Team">Team</option>
            <option value="BugsOpened">Bugs Opened</option>
            <option value="ProductionIncidents">Production Incidents
            </option>
    </select>
    X-Axis:
    <select id="xval" onChange="getFormData()">
            <option value="TotalPoints">Total Points</option>
            <option value="TotalDevs">Total Devs</option>
            <option value="Team">Team</option>
            <option value="BugsOpened">Bugs Opened</option>
            <option value="ProductionIncidents">Production Incidents
            </option>
    </select>
</form>
```

Retrieving Form Data

The last bit of functionality left is to code the getFormData() function. This function
pulls out the selected options from both select elements and use those values to pass in
to setChartDots()—after calling removeDots(), of course.

```
function getFormData(){
    var xEl = document.getElementById("xval")
    var yEl = document.getElementById("yval")
    var x = xEl.options[xEl.selectedIndex].value
    var y = yEl.options[yEl.selectedIndex].value
    removeDots()
    setChartDots(x,y)
}
```

Great!

Using the Visualization

The complete source code should look like the following:

```html
<!DOCTYPE html>
<html>
  <head>
          <meta charset="utf-8">
    <title></title>
<style>
body {
  font: 10px sans-serif;
}
.axis path,
.axis line {
  fill: none;
  stroke: #000;
  shape-rendering: crispEdges;
}
.dot {
  stroke: #000;
}
</style>
</head>
<body>
        <form>
                Y-Axis:
                <select id="yval" onChange="getFormData()">
                        <option value="TotalPoints">Total Points</option>
                        <option value="TotalDevs">Total Devs</option>
                        <option value="Team">Team</option>
                        <option value="BugsOpened">Bugs Opened</option>
                        <option value="ProductionIncidents">Production
                        Incidents</option>
                </select>
```

```
            X-Axis:
            <select id="xval" onChange="getFormData()">
                    <option value="TotalPoints">Total Points</option>
                    <option value="TotalDevs">Total Devs</option>
                    <option value="Team">Team</option>
                    <option value="BugsOpened">Bugs Opened</option>
                    <option value="ProductionIncidents">Production
                    Incidents</option>
            </select>
    </form>
<script src="d3.v3.js"></script>
<script>
var margin = {top: 20, right: 20, bottom: 30, left: 40},
    width = 960 - margin.left - margin.right,
    height = 500 - margin.top - margin.bottom,
        chartData;
var x = d3.scale.linear()
    .range([0, width]);
var y = d3.scale.linear()
    .range([height, 0]);
var xAxis = d3.svg.axis()
    .scale(x)
    .orient("bottom");
var yAxis = d3.svg.axis()
    .scale(y)
    .orient("left");
var svg = d3.select("body").append("svg")
    .attr("width", width + margin.left + margin.right)
    .attr("height", height + margin.top + margin.bottom)
  .append("g")
    .attr("transform", "translate(" + margin.left + "," + margin.top + ")");
    svg.append("g")
        .attr("class", "x axis")
        .attr("transform", "translate(0," + height + ")")
        .call(xAxis)
```

```
 svg.append("g")
     .attr("class", "y axis")
     .call(yAxis)
function getFormData(){
       var xEl = document.getElementById("xval")
       var yEl = document.getElementById("yval")
       var x = xEl.options[xEl.selectedIndex].value
       var y = yEl.options[yEl.selectedIndex].value
       removeDots()
       setChartDots(x,y)
}
function removeDots(){
     svg.selectAll("circle")
         .transition()
             .duration(0)
             .remove()
     svg.selectAll(".axis")
             .transition()
             .duration(0)
             .remove()
 }
function setChartDots(xval, yval){
     setTimeout(function() {
       x.domain(d3.extent(chartData, function(d) { return d[xval];}));
       y.domain(d3.extent(chartData, function(d) { return d[yval];}));
       svg.selectAll(".dot")
           .data(chartData)
         .enter().append("circle")
           .attr("class", "dot")
           .attr("r", 3)
           .attr("cx", function(d) { return x(d[xval]);})
           .attr("cy", function(d) { return y(d[yval]);})
           .style("fill", "#CCCCCC");
            svg.append("g")
               .attr("class", "axis")
```

```
                .attr("transform", "translate(0," + height + ")")
                .call(xAxis)
            svg.append("g")
                .attr("class", "axis")
                .call(yAxis)
        }, 300);
    }
d3.csv("teamvelocity.txt", function(error, data) {
        chartData = data;
        removeDots()
        setChartDots("TotalDevs", "TotalPoints")
});
</script>
</body>
</html>
```

And it should create the interactive visualization shown in Figure 8-13.

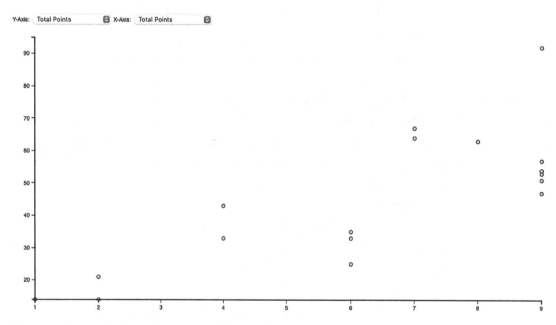

Figure 8-13. *Interactive scatter plot with D3*

Summary

This chapter looked at correlations between the speed at which a team moves and the opening of bugs and production issues. There is naturally a positive correlation between these data points: when we make new things, we create new opportunities for those new things and existing things to break.

Of course, that doesn't mean that we should stop making new things, even if for some reason our business units and our very industries would allow it. It means that we need to find balance between making new things and nurturing and maintaining the things that we already have. This is exactly what we will look at in the next chapter.

CHAPTER 9

Visualizing the Balance of Delivery and Quality with Parallel Coordinates

The last chapter looked at using scatter plots to identify relationships between sets of data. It discussed the different types of relationships that could exist between data sets, such as positive and negative correlation. We couched this idea in the premise of team dynamics: Do you see any correlation between the amount of people on a team and the amount of work that the team can complete, or between the amount of work completed and the number of defects generated?

In this chapter, we tie together the key concepts that we have been talking about: visualizing, team feature work, defects, and production incidents. We will tie them together using a data visualization called parallel coordinates to show the balance between these efforts.

What Are Parallel Coordinate Charts?

Parallel coordinate charts are a visualization that consists of N amount of vertical axes, each representing a unique data set, with lines drawn across the axes. The lines show the relationship between the axes, much like scatter plots, and the patterns that the lines form indicate the relationship. We can also gather details about the relationships between the axes when we see a clustering of lines. Let's take a look at this using the chart in Figure 9-1 as an example.

© Tom Barker, Jon Westfall 2022
T. Barker and J. Westfall, *Pro Data Visualization Using R and JavaScript*,
https://doi.org/10.1007/978-1-4842-7202-2_9

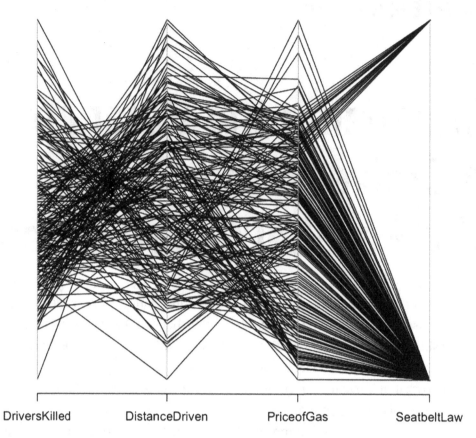

Figure 9-1. *Parallel coordinates for Seatbelts data set*

I constructed the chart in Figure 9-1 from the data set Seatbelts that comes built into R. To see a breakdown of the data set, type **?Seatbelts** at the R command line. I extracted a subset of the columns available to better highlight the relationships in the data:

```
cardeaths <- data.frame(Seatbelts[,1], Seatbelts[,5], Seatbelts[,6],
Seatbelts[,8])
colnames(cardeaths) <- c("DriversKilled", "DistanceDriven", "PriceofGas",
"SeatbeltLaw")
```

The data set represents the number of drivers killed in car accidents in Great Britain before and after it became compulsory to wear seat belts. The axes represent the number of drivers killed, the distance driven, the cost of gas at the time, and whether there was a seat belt law in place.

There are a number of useful ways to look at parallel coordinates. If we look at the lines between a single pair of axes, we can see the relationships between those data sets. For example, if we look at the relationship between the price of gas and the seat belt law, we can see that the price of gas is constrained pretty tightly for when the seat belt law was in place, but covered a large range of prices for when the seat belt law was not in place (i.e., a lot of disparate lines converge on the point that represents the time before the law, and a narrow band of lines converge on the time after the law was passed). This relationship could imply many different things, but because I know the data, I know it's because we just have a much smaller sample size for deaths after the law was put in place: 14 years' worth of data before the seat belt law, but only 2 years' worth of data after the seat belt law.

We can also trace lines across all the axes to see how each of the axes relates. This is difficult to do with all the lines of the same color, but when we change the color and shading of the lines, we can more easily see the patterns across the chart. Let's take the existing chart and assign colors to the lines (the results are displayed in Figure 9-2; also, you'll need to install the MASS package if you don't have it already):

```
library(MASS)
parcoord(cardeaths, col=rainbow(length(cardeaths[,1])), var.label=TRUE)
```

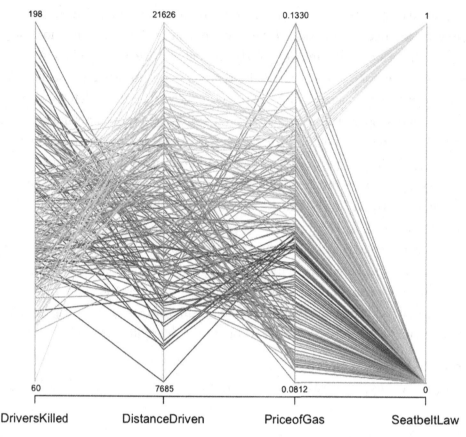

Figure 9-2. *Parallel coordinates for Seatbelts data set, with each line a different shade of gray*

Note   You need to import the MASS library to use the `parcoord()` function.

Figure 9-2 begins to show the patterns that exist in the data. The lines that have the lowest number of deaths also have the most distances driven and mainly fall into the point in time after the seat belt law was enacted. Again, note that we do have a much smaller sample size available for post–seat belt law than we do for pre–seat belt law, but you can see how it becomes useful and telling to be able to trace the interconnectedness of these data points.

History of Parallel Coordinate Plots

The idea of using parallel coordinates on vertical axes was invented in 1885 by Maurice d'Ocagne when he created the nomograph and the field of nomography. Nomographs are tools to calculate values across mathematical rules. The classic example of a nomograph still in use today is the line on a thermometer that shows values in both Fahrenheit and Celsius. Or think of rulers that show values in inches on one side and centimeters on the other.

Note   Ron Doerfler has written an extensive thesis on nomography available here: `http://myreckonings.com/wordpress/2008/01/09/the-art-of-nomography-i-geometric-design/`. Doerfler also hosts a site called modern nomograms (`www.myreckonings.com/modernnomograms/`) that "offers eye-catching and useful graphical calculators uniquely designed for today's applications."

You can see examples of modern nomograms, courtesy of Ron Doerfler, in Figures 9-3 and 9-4.

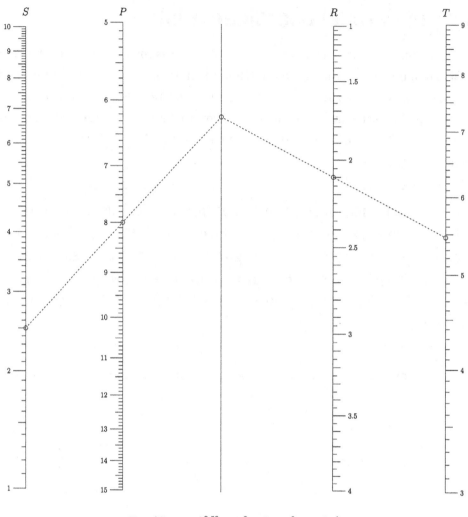

$$T = (S + 0.64)^{0.58} \times P^2 \times (1.4R^3 + 9.8)^{-1}$$

Figure 9-3. *Nomogram demonstrating the conversion of values between the functions S, P, R, and T, the basis of the sequential probability ratio test*

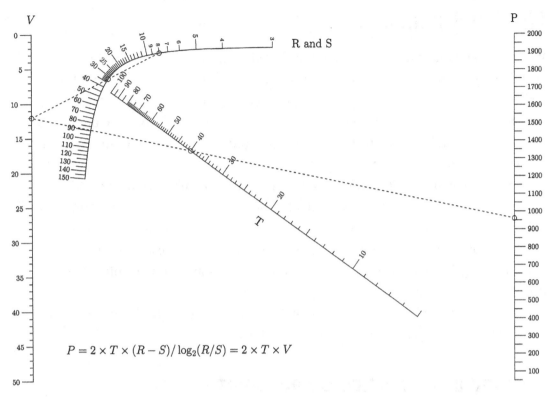

Figure 9-4. Curved scale nomogram, courtesy of Ron Doerfler, Leif Roschier, and Joe Marasco

---

Note   The term *parallel coordinates* and the concept that it represents were popularized and rediscovered by Alfred Inselberg while studying at the University of Illinois. Dr. Inselberg is currently a professor at Tel Aviv University and a Senior Fellow at the San Diego Supercomputing Center. Dr. Inselberg has also published a book on the subject, *Parallel Coordinates: Visual Multidimensional Geometry and Its Applications* (Springer, 2009). He has also published a dissertation on how to effectively read and use parallel coordinates, entitled "Multidimensional Detective," available from the IEEE.

Finding Balance

We understand that parallel coordinates are used to visualize the relationship between multiple variables, but how does that apply to what we have been talking about so far in this book? So far, we discussed quantifying and visualizing the defect backlog, the sources of the production incidents, and even the amount of work that our teams commit to. Arguably, balancing these aspects of product development can be one of the most challenging activities that a team does.

With each iteration, either formal or informal, team members have to decide how much effort they should put toward each of these concerns: working on new features, fixing bugs on existing features, and addressing production incidents from direct feedback from users. And these are just a sampling of the nuances that every product team must juggle; they also may have to factor in time to spend on technical debt or updating infrastructure.

We can use parallel coordinates to visualize this balance, both for documentation and as a tool for analysis when starting new sprints.

Creating a Parallel Coordinate Chart

There are several different approaches to creating a parallel coordinate chart. Using the data from the previous chapter, we could look at the running totals per iteration. Recall that the data was a total of points committed to per iteration, as well as a snapshot of how many bugs and production incidents were in each team's backlog, how many new bugs were opened during the iteration, and how many members there were on the team. The data looked much like this:

Sprint	TotalPoints	TotalDevs	Team	BugBacklog	BugsOpened	ProductionIncidents
1	12.10	25	6 Gold	125	10	1
2	12.20	42	8 Gold	135	30	3
3	12.30	45	8 Gold	150	25	2
4	12.40	43	8 Gold	149	23	3
5	12.50	32	6 Gold	155	24	1
6	12.60	43	8 Gold	140	22	4
7	12.70	35	7 Gold	132	9	1

...

To make use of this data, we can read it in to R, just as we did in the last chapter:

```
tvFile <- "/Applications/MAMP/htdocs/teamvelocity.txt"
teamvelocity <- read.table(tvFile, sep=",", header=TRUE)
```

We then can create a new data frame with all the columns from the `teamvelocity` variable except the Team column. That column is a string, and the R `parcoord()` function, which we use in this example, throws an error if we include strings in the object that we pass in to it. Team information wouldn't make sense in this context, either. The lines that will be drawn in the chart will be representative of our teams:

```
t<- data.frame(teamvelocity$Sprint, teamvelocity$TotalPoints,
teamvelocity$TotalDevs, teamvelocity$BugBacklog, teamvelocity$BugsOpened,
teamvelocity$ProductionIncidents)
colnames(t) <- c("sprint", "points", "devs", "total bugs", "new bugs",
"prod incidents")
```

We pass the new object into the `parcoord()` function. We also pass the `rainbow()` function into the `color` parameter, as well as set the `var.label` parameter to `true`, to make the upper and lower boundaries of each axis visible on the chart:

```
parcoord(t, col=rainbow(length(t[,1])), var.label=TRUE)
```

This code produces the visualization shown in Figure 9-5.

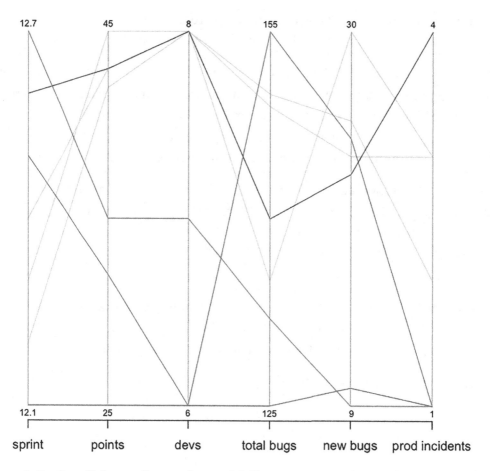

Figure 9-5. *Parallel coordinate chart of different aspects of overall organizational metrics, including points committed to per iteration, total developers by team, total bug backlog, new bugs open, and production incidents*

Figure 9-5 presents some interesting stories for us. We can see that some teams in our data set create more bugs as they take on more points' worth of work. Other teams have a large bug backlog while not creating a large number of new bugs during each iteration, which implies that they are not closing the bugs that they do open. Some teams are more consistent than others. All contain insights that the teams can use for reflection and continual improvement. But ultimately this chart is reactive and talks around the main issues. It tells us what the effects of each sprint are on our respective backlogs, both bugs and production incidents. It also tells us how many bugs were opened during each sprint.

What the figure doesn't show is the amount of effort spent working against each backlog. To show that, we need to do a bit of prep work.

Adding in Effort

Past chapters I mentioned Greenhopper and Rally as ways to plan iterations, prioritize backlogs, and track progress on user stories. No matter the product you choose, it should provide some way to categorize or tag your user stories with metadata. Some very simple ways to accomplish this categorization without needing your software to support it include these:

- Put tagging in the title of each user story (see Figure 9-6 for an example of what this could look like in Rally). With this method, you need to sum the level of effort for each category, either manually or programmatically.

Figure 9-6. *User stories tagged by category, Defect, Feature, or Prod Incident (courtesy of Rally)*

- Nest subprojects for each delineation of effort.

However you go about creating these buckets, you should have a way to track the amount of effort spent during each sprint for your categories. To visualize this, just export it from your favorite tool into a flat file that resembles the structure shown here:

```
iteration,defect,prodincidents,features,techdebt,innovation
13.1,6,3,13,2,1
13.2,8,1,7,2,1
13.3,10,1,9,3,2
```

```
13.5,9,2,18,10,3
13.6,7,5,19,8,3
13.7,9,5,21,12,3
13.8,6,7,11,14,3
13.9,8,3,16,18,3
13.10,7,4,15,5,3
```

To begin using this data, we need to import the contents of the flat file into R. We store the data in a variable named teamEffort and pass teamEffort into the parcoord() function:

```
teFile <- "/Applications/MAMP/htdocs/teamEffort.txt"
teamEffort <- read.table(teFile, sep=",", header=TRUE)
parcoord(teamEffort, col=rainbow(length(teamEffort[,1])), var.label=TRUE,
main="Level of Effort Spent")
```

This code produces the chart shown in Figure 9-7.

Level of Effort Spent

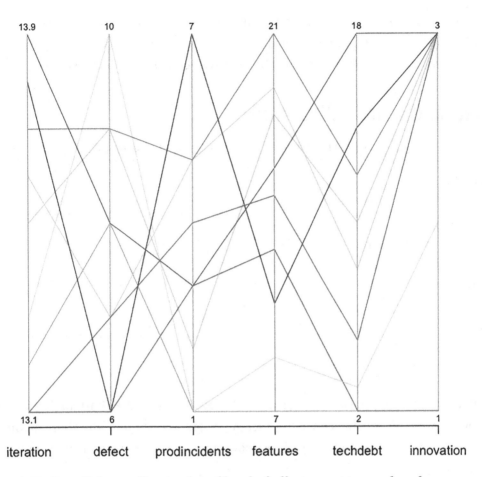

Figure 9-7. *Parallel coordinate plot of level of effort spent toward each initiative*

This chart is less about seeing relationships implied by data and more about seeing explicit levels of effort committed to each sprint. In a vacuum, these data points are meaningless, but when you look at both charts and compare the total bug backlog and total production incidents, compared with the level of effort spent addressing either, you begin to see blind spots that the team would need to address. Blind spots might be where teams that have high bug backlogs or production incident counts are not spending enough effort to address those backlogs.

Brushing Parallel Coordinate Charts with D3

The trick to reading dense parallel coordinate plots is to use a technique called brushing. Brushing fades the color or opacity of all the lines on the chart, except for the lines you want to trace across the axes. We can achieve this level of interactivity using D3.

Creating the Base Structure

Let's start by creating a new HTML file using our base HTML skeletal structure:

```
<!DOCTYPE html>
<html>
  <head>
          <meta charset="utf-8">
    <title></title>
</head>
<body>
<script src="d3.v3.js"></script>
</body>
</html>
```

We then create a new `script` tag to hold the JavaScript for the chart. In this tag, we start by creating the variables needed to set the height and width of the chart, an object to hold the margin values, an array of axes column names, and the scale object for the x object.

We also create variables to reference the D3 SVG line object, a reference to the D3 axis, and a variable named `foreground` to hold the groupings of all the paths that will be the lines drawn between axes in the chart:

```
<script>
var margin = {top: 80, right: 160, bottom: 200, left: 160},
     width = 1280 - margin.left - margin.right,
     height = 800 - margin.top - margin.bottom,
         cols =
["iteration","defect","prodincidents","features","techdebt","innovation"]
var x = d3.scale.ordinal().domain(cols).rangePoints([0, width]),
    y = {};
```

```
var line = d3.svg.line(),
    axis = d3.svg.axis().orient("left"),
    foreground;
</script>
```

We draw the SVG element to the page and store it in a variable that we name **svg**:

```
var svg = d3.select("body").append("svg")
    .attr("width", width + margin.left + margin.right)
    .attr("height", height + margin.top + margin.bottom)
  .append("g")
    .attr("transform", "translate(" + margin.left + "," + margin.top +
")");
We use d3.csv to load in the teameffort.txt flat file:
d3.csv("teameffort.txt", function(error, data) {
}
```

So far, we're following the same format as in previous chapters: lay out variables at the top, create the SVG element, and load in the data; most of the data-dependent logic happens in the anonymous function that fires when the data has been loaded.

For parallel coordinates, this process changes a bit right here because we need to create y-axes for each column in our data.

Creating a Y-Axis for Each Column

To create a y-axis for each column, we have to loop through the array that holds the column names, convert the contents of each column to explicitly be numbers, create an index in the y variable for each column, and create a D3 `scale` object for each column:

```
cols.forEach(function(d) {
        //convert to numbers
        data.forEach(function(p) { p[d] = +p[d]; });
        //create y scale for each column
        y[d] = d3.scale.linear()
                .domain(d3.extent(data, function(p) { return p[d]; }))
                .range([height, 0]);
});
```

Drawing the Lines

We need to draw the lines that will traverse each axis, so we create an SVG grouping to aggregate and hold all the lines. We assign the foreground class to the grouping (doing so is important because we will handle the brushing of the lines via CSS):

```
foreground = svg.append("g")
        .attr("class", "foreground")
```

We append SVG paths to this grouping. We attach the data to the paths, set the color of the paths to randomly generated colors, and stub out mouseover and mouseout event handlers. We also set the d attribute of the paths to a function that we will create called path().

We'll come back to those event handlers in a minute.

```
foreground = svg.append("g")
    .attr("class", "foreground")
  .selectAll("path")
    .data(data)
  .enter().append("path")
.attr("stroke", function(){return "#" + Math.floor(Math.random()*16777215).
toString(16);})
    .attr("d", path)
    .attr("width", 16)
        .on("mouseover", function(d){
        })
        .on("mouseout", function(d){
        })
```

Let's flesh out the path() function. In this function, we accept a parameter named d, which will be an index of the data variable. The function returns a mapping of the path coordinates with the x and y scales.

```
function path(d) {
    return line(cols.map(function(p) { return [x(p), y[p](d[p])]; }));
}
```

The path() function returns data that looks much like the following—a multidimensional array with each index and array consisting of two coordinate values:

```
[[0, 520], [192, 297.14285714285717], [384, 346.6666666666667], [576, 312],
[768, 491.1111111111111], [960, 520]]
```

Fading the Lines

Let's take a step back for a second. To handle the brushing, we need to create a style rule to fade the opacity of the lines. So let's return to the head section of the page and create a style tag and some style rules.

We set path.fade as the selector and set the stroke-opacity to 4%. While we're at it, we also set body font styles and path styles.

```
<style>
body {
  font: 15px sans-serif;
  font-weight:normal;
}
path{
  fill: none;
  shape-rendering: geometricPrecision;
  stroke-width:1;
}
path.fade {
  stroke: #000;
  stroke-opacity: .04;
}
</style>
```

Let's return to the stubbed out event handlers. D3 provides a function called classed() that allows us to add classes to selections. In the mouseover handler, we use the classed() function to apply the fade style that we just created to every path in the foreground. It fades out each line. We'll next target the current selection, using d3. select(this) and classed() to turn off the fade styling.

In the mouseout handler, we turn off the fade style:

```
foreground = svg.append("g")
    .attr("class", "foreground")
  .selectAll("path")
    .data(data)
  .enter().append("path")
 .attr("stroke", function(){return "#" + Math.floor(Math.
 random()*16777215).toString(16);})
    .attr("d", path)
    .attr("width", 16)
      .on("mouseover", function(d){
              foreground.classed("fade",true)
              d3.select(this).classed("fade", false)
      })
      .on("mouseout", function(d){
              foreground.classed("fade",false)
      })
```

Creating the Axes

Finally, we need to create the axes:

```
var g = svg.selectAll(".column")
              .data(cols)
            .enter().append("svg:g")
              .attr("class", "column")
                  .attr("stroke", "#000000")
              .attr("transform", function(d) { return "translate(" + x(d)
              + ")"; })
          // Add an axis and title.
          g.append("g")
              .attr("class", "axis")
              .each(function(d) { d3.select(this).call(axis.scale(y[d])); })
            .append("svg:text")
              .attr("text-anchor", "middle")
```

```
            .attr("y", -19)
            .text(String);
```

Our complete code is as follows:

```
<!DOCTYPE html>
<html>
  <head>
            <meta charset="utf-8">
    <title></title>
<style>
body {
  font: 15px sans-serif;
  font-weight:normal;
}
path{
  fill: none;
  shape-rendering: geometricPrecision;
  stroke-width:1;
}
path.fade {
  stroke: #000;
  stroke-opacity: .04;
}
</style>
</head>
<body>
<script src="d3.v3.js"></script>
<script>
var margin = {top: 80, right: 160, bottom: 200, left: 160},
    width = 1280 - margin.left - margin.right,
    height = 800 - margin.top - margin.bottom,
        cols = ["iteration","defect","prodincidents","features",
"techdebt","innovation"]
var x = d3.scale.ordinal().domain(cols).rangePoints([0, width]),
    y = {};
```

```
var line = d3.svg.line(),
    axis = d3.svg.axis().orient("left"),
    foreground;
var svg = d3.select("body").append("svg")
    .attr("width", width + margin.left + margin.right)
    .attr("height", height + margin.top + margin.bottom)
  .append("g")
    .attr("transform", "translate(" + margin.left + "," +
    margin.top + ")");
d3.csv("teameffort.txt", function(error, data) {
        cols.forEach(function(d) {
                //convert to numbers
            data.forEach(function(p) { p[d] = +p[d]; });
            y[d] = d3.scale.linear()
                .domain(d3.extent(data, function(p) { return p[d]; }))
                .range([height, 0]);
                });
            foreground = svg.append("g")
                .attr("class", "foreground")
              .selectAll("path")
                .data(data)
              .enter().append("path")
            .attr("stroke", function(){return "#" + Math.floor(Math.
            random()*16777215).toString(16);})
                .attr("d", path)
                .attr("width", 16)
                    .on("mouseover", function(d){
                            foreground.classed("fade",true)
                            d3.select(this).classed("fade", false)
                    })
                    .on("mouseout", function(d){
                            foreground.classed("fade",false)
                    })
```

```
        var g = svg.selectAll(".column")
            .data(cols)
          .enter().append("svg:g")
            .attr("class", "column")
                .attr("stroke", "#000000")
            .attr("transform", function(d) { return "translate(" + x(d)
            + ")"; })
        // Add an axis and title.
        g.append("g")
            .attr("class", "axis")
            .each(function(d) { d3.select(this).call(axis.scale(y[d]));
            })
          .append("svg:text")
            .attr("text-anchor", "middle")
            .attr("y", -19)
            .text(String);
            function path(d) {
                return line(cols.map(function(p) { return [x(p), y[p]
                (d[p])]; }));
              }
        });
</script>
</body>
</html>
```

This code produces the chart shown in Figure 9-8.

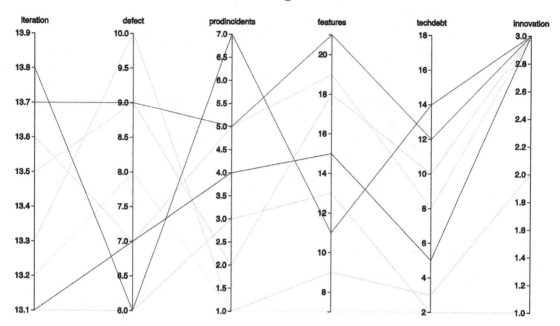

Figure 9-8. *Parallel coordinate chart created in D3*

If we roll over any of the lines, we see the brushing effect shown in Figure 9-9, in which the opacity of all the lines, except the one currently moused over, is scaled back.

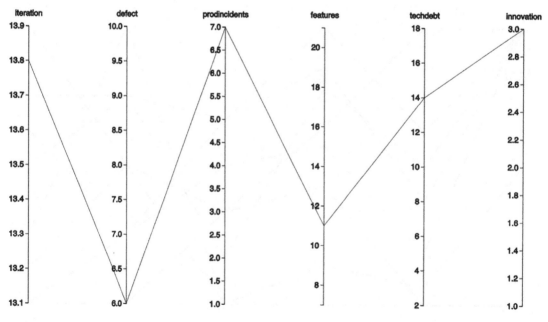

Figure 9-9. *Parallel coordinate chart with interactive brushing*

Summary

This chapter looked at parallel coordinate charts. You got a taste of their history—how they came about originally in the form of nomograms used to show value conversions. You looked at their practical application in the context of visualizing how teams balance the different aspects of product development in the course of an iteration.

Parallel coordinates are the last visualization type covered in this book, but it is far from the last type of visualization out there. And this book is far from the last word on the subject. Something that I tell my students at the end of each semester is that I hope they will continue to use what they have learned in my class. Only by using the language or subject that was covered, by continually playing with it, exploring it, and testing the boundaries of it will students incorporate this new tool into their own skillset. Otherwise, if they leave the class (or, in this case, close the book) and not think about the subject for a good while, they will probably forget much of what we went over.

If you are a developer or technical leader, I hope that you read this book and were inspired to begin tracking your own data. This was just a small sampling of things that you can track. You can instrument your code to track performance metrics, as covered in my book *Pro JavaScript Performance: Monitoring and Visualization*, or you can use tools such as Splunk to create dashboards to visualize usage data and error rates. You can tap right into the source code repository database to get such metrics as what times and days of the week have the most commit activity to avoid scheduling meetings during those times.

The point of all this data tracking is self-improvement—to establish baselines of where you currently are and track progress toward where you want to be, to constantly refine your craft, and excel at what you do.

Index

A

Access logs
  Apache documentation, 122
  data map, R
    displaying regional data, 144–146
    geographic data, 137, 139, 140
    latitude/longitude, 141, 143
  definition, 121
  distributing visualization, 146, 147, 149
  documentation, 122
  parsing
    control logic, 134–136
    geolocation IP, 129–131
    output fields, 132
    parse log file, 125–128
    process, 123
    read, 124
aggregate() function, 199
Agile development, 224
Anonymous function, 58, 69, 100, 107,
        108, 138, 144, 167
append() function, 98
Application programming
        interface (API), 20, 90
apply() function, 57, 76, 144
as.Date() function, 155
as.matrix() function, 53
attr() function, 97, 100, 107
axis() function, 161

B

Bar chart, 186
  bugs, 187
  D3, 203
  groups, 191, 192
  horizontal, 204
  plot data, 197, 199
  production incidents, 201
  stacked, 188, 191
  standard, 186, 187
Bubble charts, 222, 223, 227–229
Bug-tracking software, 154, 178,
        228, 229

C

Cascading Style Sheet (CSS), 90, 206
chartData variable, 238
Cholera map, 9, 10, 117
Cite sources, 32
classed() function, 263
colnames() function, 45
Comprehensive R Archive Network
        (CRAN), 34, 35
Correlation analysis, 225
  bubble chart, 227
  scatter plot, 226
createPlaylist() function, 67
cumsum() function, 159

T. Barker and J. Westfall, *Pro Data Visualization Using R and JavaScript*,
https://doi.org/10.1007/978-1-4842-7202-2

Printed in the United States
by Baker & Taylor Publisher Services